Doing Cultural Geography

DOING HUMAN GEOGRAPHY

The 'Doing Human Geography' series places the acquisition of skills at the core of learning, and each volume is written with the Geography Benchmark statement in mind. The emphasis on active learning underlines the value of a range of theoretical positions in economic, political, social and cultural geography, and allows students to get the most out of geography fieldwork.

Titles in the series.

Doing Economic Geography
Edited by Simon Leonard

Doing Political Geography
Edited by Giles Mohan

Doing Social Geography
Edited by Pamela Shurmer-Smith

Doing Fieldwork Geography
Edited by Carol Ekinsmyth

Doing Cultural Geography

edited by
PAMELA SHURMER-SMITH

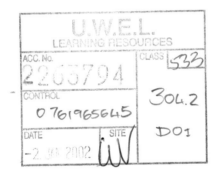

SAGE Publications
London • Thousand Oaks • New Delhi

First published 2002

SAGE Publications Ltd
6 Bonhill Street
London EC2A 4PU

SAGE Publications Inc.
2455 Teller Road
Thousand Oaks, California 91320

SAGE Publications India Pvt Ltd
32, M-Block Market
Greater Kailash – I
New Delhi 110 048

British Library Cataloguing in Publication data

A catalogue record for this book is available from the British Library

ISBN 0 7619 6564 5
ISBN 0 7619 6563 3 (pbk)

Library of Congress Control Number 2001 135325

Typeset by Photoprint, Torquay, Devon
Printed in Great Britain by The Cromwell Press Ltd,
Trowbridge, Wiltshire

Doing Cultural Geography

edited by
PAMELA SHURMER-SMITH

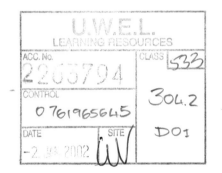

SAGE Publications
London • Thousand Oaks • New Delhi

First published 2002

SAGE Publications Ltd
6 Bonhill Street
London EC2A 4PU

SAGE Publications Inc.
2455 Teller Road
Thousand Oaks, California 91320

SAGE Publications India Pvt Ltd
32, M-Block Market
Greater Kailash – I
New Delhi 110 048

British Library Cataloguing in Publication data

A catalogue record for this book is available from the British Library

ISBN 0 7619 6564 5
ISBN 0 7619 6563 3 (pbk)

Library of Congress Control Number 2001 135325

Typeset by Photoprint, Torquay, Devon
Printed in Great Britain by The Cromwell Press Ltd,
Trowbridge, Wiltshire

Contents

1 Introduction

Pamela Shurmer-Smith

As people have become increasingly aware of the radical changes inherent in late modern society, there has been a growing desire to find new ways of thinking in order to reach new modes of understanding. The oil crisis of 1973 and the ensuing new wave of neo-liberal 'globalization' caused fundamental changes world-wide. These changes were not just in the economic and political order, but permeated into the recesses of ordinary people's lives. Not only in the universities, but also in the media and in private encounters, virtually everyone, everywhere became increasingly conscious of the problem of creating meaning in situations in which so many of the parameters of economic, political and social life had shifted.

It is now taken for granted that the implications of a globally integrated system of production and consumption involve everyone, everywhere: industrial workers, peasant farmers, landless labourers, retired people, traders, professionals, students, people bringing up children and those trying to find work. Men and women contemplated and tried to make sense of such things as insecure employment, changed levels and modes of consumption, the obsolescence of hard won knowledge, the sidelining of old moral certainties. They speculated about whether these things were the cause or the outcome of changed relationships and power structures. They worried about changing families and sexual relationships, new expectations regarding gender and age, different ways of managing power and authority, unaccustomed notions of justice, liability, accountability and entitlement, the decline in religious observance in some places and virulent new forms of anti-secularism in others.

It was in this climate of general consciousness of the problem of finding meaning and value that the so-called 'cultural turn' occurred amongst intellectuals in all of the social sciences, and not only in the West. Indeed, there came to be some unease about the use of the term 'science' when studying society. By the 'cultural turn', it was implied that the accumulations of ways of seeing, means of communicating, constructions of value, senses of identity should be taken as important in their own right, rather than just a by-product of economic formations. (It still remains to be seen whether this new orthodoxy is correct.) Suddenly 'culture' became intellectually fashionable as a starting point for interpretation, whereas it had hitherto been seen as lacking in rigour.

Geography was no exception to this and gradually, through the 1980s, all of the sub-disciplines of human geography came to be conscious of the 'cultural' dimensions of their field of study: economic geographers 'discovered' embeddedness of local economies in local social practices; political geographers became aware of new nationalisms and notions of identity in boundary formation and exclusion; urban geographers turned their attention to lifestyle and they became enthusiastic about cultural regeneration of cities; the countryside was rethought as a cultural construction, as was nature itself; retail geographers became enthusiastic about sites of consumption, as opposed to patterns of distribution. Difference and differentiation shifted to centre stage, and yet no one was very clear about what 'culture' might be. In 1988 the Social Geography Study Group of the Institute of British Geographers launched its 'cultural initiative' and simultaneously changed its name to the Social *and Cultural* Geography Study Group. The following year Peter Jackson's (1989) *Maps of Meaning* irrevocably changed the way in which geographers looked at issues of identity. The sub-discipline has taken off from here in its many different guises.

> The change in title [of the Social Geography Study Group to include Cultural Geography] is an entirely welcome event for someone like myself who has always believed that human geography should celebrate the cultural diversity of our world and pay attention to the ways in which human beliefs, values and ideals continuously shape its landscapes. It is a change which signals a profound and, in some respects, an overdue change in geographical philosophy and methodology.
>
> (Cosgrove 1988, quoted in Philo 1991 p. 1)

Trevor Barnes' (1996) *Logics of Dislocation* is a stimulating study of the ways in which academic work cannot justifiably claim to rise above the values and modes of communication of the people who produce it. He demonstrates the significance of metaphor in economic geography, probably the branch of human geography which has been the most assiduous in attempting to avoid cultural 'bias' by adopting the supposedly universal, and therefore culture-free, language of mathematics. Barnes argues that the language of economic geography is as cultural as any other and just as subject to deconstruction. It is a book one could hardly have conceived of before the 'cultural turn'.

One of the most influential texts in American cultural geography defines culture quite unambiguously as 'a total way of life held in common by a group of people' (Jordan and Rountree 1982 p. 4). That might seem straightforward enough but, when one thinks about it, it is meaningless.

Which of us shares *every* aspect of our life with even one other person, let alone a group, presumably a group large enough to constitute a society? What is meant by 'a total way of life'? Does it mean that everyone has to act and think the same way, or does it mean that some, unspecified, degree of (recognized) difference is permitted? How does 'a way of life held in common' cope with change: does everyone have to change in the same way at the same time? Sperber suggests that we should see culture as a 'polythetic term' (1996 p. 17), meaning a condition in which there is 'a set of features such that none of them is necessary, but any large enough subset of them is sufficient for something to fall under the term' (he simplifies a little by suggesting one thinks of this as a 'family resemblance'). The very idea of culture as the *possession* of bounded groups of people is divisive and dangerous as it underpins unnecessary oppositions and enmities.

Although I have been identified with cultural geography for more than a decade, I do not believe (and never have believed) that there is any such *thing* as culture. Like Don Mitchell (1995; 2000) I have insisted (Shurmer-Smith and Hannam 1994) that culture is practised, not owned. It is what people *do*, not what they have, and they keep doing different things in different ways, with different other people all of the time. Wendy James puts it concisely when she maintains that we should think of culture as being 'adverbial' rather than 'nominal' (1996 p. 106). Culture, then, is the communicating, sense-making, sharing, evaluating, wondering, reinforcing, experimenting qualifier of what people *do*. Richards likens it to music, in which the performance is everything, where 'manifestly, the purpose . . . is not to arrive at the end' (1996 p. 126); but, for me, it is like music only if we are willing to include solo improvisation and particularly 'jamming', playing in creative dialogue with others, who understand what one is doing, recognize the underlying theme, but are still constantly surprised. I prefer this metaphor to the one which underlies Sperber's view that 'Culture is made up, first and foremost, of contagious ideas' (1996 p. 1), which leads him to think in terms of an epidemiology.

Cultural geography, then, becomes the field of study which concentrates upon the ways in which space, place and the environment participate in an unfolding dialogue of meaning. This includes thinking about how geographical phenomena are shaped, worked and apportioned according to ideology; how they are used when people form and express their relationships and ideas, including their sense of who they are. It also includes the ways in which place, space and environment are perceived and represented, how they are depicted in the arts, folklore and media and how these artistic uses feed back into the practical. People often think of 'culture' and 'tradition' as being synonymous, but they are not; much of culture is new and conscious of its newness (and much of tradition only pretends to be old, but that is another matter). Along with many other students of 'culture', I am convinced that power is always involved in human constructions, communications and representations, and that one cannot, practically or conceptually, separate the cultural from the political. Different ways of behaving and making sense not only clash and compete but also contribute to the formation of each other in that conflict.

Terry Eagleton, a leading cultural theorist, concludes his most recent book by saying that 'culture has assumed a new political importance. But it has grown, at the same time, immodest and overweening. It is time, while acknowledging its significance, to put it back in its place' (2000 p. 131). I suggested in 1994 (Shurmer-Smith and Hannam) that the excitement with culture might rapidly evaporate, and Don Mitchell (1999) bravely announced the demise of cultural geography shortly before the publication of his excellent (2000) contribution to the field. It is not that any of us has a death wish; it is much more that those of us who believe that culture is lived, not owned, need constantly to fight against those who reify (even deify) it.

This book will strive to put culture in its place (and space, and environment). The aim is to demystify culture, maintaining that it has nothing to do with muses or eternal spirits, though one might concede space to Lorca's *duende*, that passionate force which is generated when an artist is in full flow (Lorca 1975). Instead it intends to show that what matters is the performance of ordinary material beings, in their relationships with each other and with the world about them. Sometimes this performance is deliberately mystified or made artificially exclusive; at other times it can be mundane, or even vulgar and offensive. Although culture is not some essential and external force, this does not imply that it is either insubstantial or unimportant, for without the ability to conceptualize and communicate we would be less than human. As a risky corollary to this, I would assert that the more people we can conceptualize alongside and communicate with, the greater our humanity. This does not imply that everyone has to become the same. Whilst the view of 'culture' as the possession or identifier of groups imposes barriers between people, a concept of culture as performance allows a movement towards innovative communication, a sort of global 'jamming'.

LEARNING VERSUS TEACHING

Given the performative view of culture I have expressed, it would be contradictory for this book to attempt to pin down a body of received wisdom relating to theory and practice in cultural geography. It would be even more inappropriate if it were to work its way round the world describing different 'cultures'. The intention is to encourage people to *do* cultural geography. *Doing* includes looking, feeling, thinking, playing, talking, writing, photographing, drawing, assembling, collecting, recording and filming as well as the more familiar reading and listening. There is no reason automatically to associate 'doing' with conducting research, although far too many academics think of that as the pre-eminent activity. *Doing* means being *active*, avoiding *being done to* (passive), so any reading and listening must be critical and part of a dialogue. If the book is successful, learning will have taken place, but it is to be hoped that every reader will have learned different things because all the contributors have been asked to curtail the desire to teach in the sense of dictating knowledge. Learning is active, being taught is passive and, just as passive bodies atrophy, so do passive minds.

The way of practising in an academic field which customarily receives the most attention is research, which is rather strange, since the majority of undergraduate students have no desire to spend their working lives in academia. This book certainly does not pretend to be yet another manual of research methods, but it does recognize that an understanding of some of the processes involved in research helps one to read academic texts in a more engaged and critical fashion. In just the same way, an awareness of theoretical debates helps one to see the point of studies which might otherwise seem arcane. Knowing how research is undertaken helps one understand how the artefacts (books, papers, lectures etc.) are crafted. Even cultural geography generates its own cultural forms, complete with mystifications and exclusions; doing it is the way to break down the defences between teachers and taught.

All of the authors believe that cultural geography (or any other subject) can be pursued actively at a number of different scales, from a few moments contemplating the meaning of the spatial elements in an advertisement to a lengthy research based monograph. We certainly do not think that cultural geography should be confined to the classroom or library; the *doing* should permeate life outside, but then should be brought back to enrich the academic endeavour. One can *do* cultural geography by being conscious of the range of geographical meanings and communications within ordinary activities, the manipulation of places and dispositions of people in spaces that only *seem* like second nature. There is spatial politics in aggressively pushing or politely standing back in a shop; in considerately making oneself as small as possible or comfortably spreading one's legs in tourist class aeroplane seats; in covering one's windows in voile or displaying one's life to the world. These sorts of things *may* be entirely individually motivated, but I'm inclined to look for meanings and social patterns.

THE STRUCTURE OF THE BOOK

This book is structured into three distinct parts, but themes will be carried through from one part to another. Although we have designed the book with a sequential logic in mind, we do not really expect most readers to start at the beginning and keep going until the end (and then stop). The logic we have used is that of a typical programme of research, starting by thinking about theory and ending up worrying about presentation. However, as has already been said, we do not assume that all of our readers are undertaking a large research project and much of what we say is addressed to people who are doing cultural geography in different ways and at different levels.

Most chapters are written by a single author and, though we often agree with each other, we are not clones. Our different takes on cultural geography demonstrate our underlying assumption about cultural activity, namely that, even when people are in close communication, purposes and meanings never absolutely converge. We have different estimations of the value of various theoretical positions, rate different authors, favour different research methods and have different ways of presenting our thoughts. But we do all subscribe to the same common endeavour and I think we all believe that if one isn't constantly vigilant, one can wind up being dragooned into some very unacceptable positions. We all believe that consciousness about value and meaning is important. None of the chapters purports to do more than start to break into the various areas of cultural geography and its associated methods. We have written assuming that our readers are just starting out on this sort of work and we have not attempted to compete with works addressed to people with more experience. We have placed our emphasis on brevity and this clearly means that subtlety has to go. We expect that readers will find some of our statements oversimplified – but the point was to encourage people to react and be encouraged to read the specialist literature.

The first part tackles the question of 'theory'. It maintains that too many professional theoreticians have had a vested interest in making theory seem difficult and have constructed it as the restricted domain of the most erudite. In truth, theory is present every time someone says, 'in my opinion . . .'. It is, however, quite important not only that one becomes conscious about one's own theoretical position, but also that one can spot where an author is coming from – and, unfortunately, this can entail some rather hard work. There is no way in which we can do justice to all the useful cultural theory in a few pages and we are resistant to trying to give potted summaries which oversimplify. The theoretical part intends to help our readers to decide where they need to read further, depending upon their own perspectives.

The second part begins by thinking about the selection of specific areas of study *through which* one can do cultural geography. We stress that the discovery of process is more important than the topic itself: almost anything can be used as the gateway into systems of meaning and value. Few things are inherently trivial if one is using them to gain access to cultural processes, even though journalists like to pick on and trivialize what they know their readers will consider not to be proper 'academic' subjects (such as clubbing, footballers as icons, food fashions). We do, however, warn that it is important to select topics which are of an appropriate scale for the project one is attempting, whether one is preparing for a class discussion or a final year thesis.

This part continues by marrying methods with theoretical position, topic and scale of project. It is important to remember that methodology means the theory of method; it is the way in which one approaches the gathering of information. There is nothing mechanical about methodology: one cannot simply pull an intact method down from the shelf, it always needs tailoring to one's own requirements and the requirements of the situation in which one is working. Given that cultural geography is concerned with trying to understand ideas and performances that are in constant flux, it is obvious that one is going to have to innovate within realms of accepted practice, rather than follow strict recipes.

The third part, on interpretation of findings, brings theory back to centre stage, for facts are not in themselves interesting and they certainly never speak for themselves. This part is itself deeply value laden since it needs to confront dangerous ideas about 'truth' and 'partiality', and especially the vexed question of subjectivity.

If interpretation is a problem, presentation is an even greater one. The penultimate chapter considers how others will access, confront and be able to respond to one's ideas and findings. Everyone who writes weighs up the words, worries about their multiple meanings, considers tone and register. How light, how serious, how much does one want to seem accessible, how much does one want to impress? But it is not just a matter of writing: we need to consider in which contexts film or video may be the best medium, and best for what purpose. We have to decide whether it is ever appropriate in an academic context to attempt an 'alternative' mode of presentation (such as poetry or plays), when one's audience may not know the conventions and thus be disempowered in critique. Presentation is always concerned with seizing power or deferring to it; one needs to recognize this, not only when making decisions about how to present one's work, but also when trying to work out how to understand what is laid before one.

I have a cartoon in which a lecturer confronts a new student with the cruel words, 'Why don't you leave now, while you still know everything?' True as it is, I have never had the heart to pin it on my board. I cannot really decide whether one is happiest when one is stupidly certain or intelligently confused, but I do know which I prefer as a way of confronting the world. What do you think?

FURTHER READING

Jackson, P. (1989) *Maps of Meaning*. London: Unwin Hyman.

Mitchell, D. (2000) *Cultural Geography: a Critical Introduction*. Oxford: Blackwell.

Shurmer-Smith, P. and Hannam, K. (1994) *Worlds of Desire, Realms of Power: a Cultural Geography*. London: Edward Arnold.

Thrift, N. (1996) *Spatial Formations*. London: Sage.

2 The trouble with theory

Pamela Shurmer-Smith

The main trouble with theory is that people will insist on making such a song and dance about it, but (as Alain de Botton 2000 demonstrated with his improbable best-seller on the uses of philosophy) the intention of theory is to make thinking easier, not more difficult. Trying to commit lists of names and salient points of different theoretical positions to memory usually results in nothing but confusion and failure; the published tables of swot notes one sometimes encounters often make things *look* simple, whilst rendering understanding near impossible (Kitchen and Tate 2000 pp. 20–2). One simply doesn't use theory like that.

The way to understand theory is to start by recognizing that everyone is (always and already) using it to construct meaning. All ways of looking at things – ideological positions, values and morals, notions of relevance and utility, all expectations, ways of classifying – are theoretical. 'Common sense' is inherently theoretical, as one invariably notices when confronting a version of 'common sense' that one cannot subscribe to. Things can only be thought by becoming phenomena, that is, by being capable of being sensed: as soon as we see, hear, feel, smell something we theorize it, explain it to ourselves (even if only to say that it is not important). This applies in everyday life and in the natural sciences just as much as in the arts, humanities and social sciences. Throughout the world, there are people who specialize in bringing latent theory to the surface; obviously academics come into this category, but so do religious specialists, artists and politicians. All of these to some extent reify theory – make it into a thing in itself, lever it apart from its ordinary implication within life. Though it is tempting to think that they behave like this in order to distinguish themselves from normal people who just get on with the mundane tasks of life, they usefully draw attention to how different ways of looking at the world privilege and disadvantage various human positions.

Mapping theoretical perspectives is not at all interesting until one has begun to appreciate that different ways of looking at the world result in different knowledges. Once one has this sense of difference, the labels (the -ologies and -isms) become convenient distinguishing tools in one's thinking. Starting the other way round, with a heap of labels and the task of finding examples to fit them, seems like a rather futile exercise to most people.

It certainly helps if one is willing to look at one's own 'of course' way of thinking and to curb any tendency to start from a judgemental position when looking at that of other people. Most of us argue our own corner when we confront an ideological position different from our own. Many of the books I possessed as a student still bear aggressive pencil marks in the margin along the lines of 'Stupid!', 'Pretentious rubbish', 'Typical racist!' and, though this was a starting point in distinguishing between ideas, it did not get me very far in *understanding* the people I was casting out beyond the pale of reasonable thought. I still do not espouse views I dislike, but I have learned to be fascinated by them and to try looking from different perspectives. This is not an exercise in tolerance. The aim is to discover *why* it is satisfying and comfortable for some people to think in one way and some in others; this inevitably means becoming conscious of the politics inherent in theory.

There it is no outside position when theorizing culture, not even when one is trying to understand the ideas and values of 'others'. Culture is inherently theoretical, and to comment upon it is to theorize about theory. It requires that one becomes self-reflexive, conscious of one's own view-point when trying to understand viewpoints in general.

> The definition of 'Culture' as a whole way of life is perhaps the most interesting since this can be read in a great many ways. For instance, references are often made to 'Indian culture' or 'Hindu culture' or, more plausibly, 'Brahminical culture' or 'upper class culture'. The two latter claims are more plausible because members of the same consolidated caste or class tend to share broad parameters of a certain culture. But usages where culture is identified with a nation state or religion tend to obfuscate matters and they often conceal a demonstrable degree of aggression behind benign sounding cultural invocation.
>
> (Ahmad 1997a p. 77)

In opposition to the increasing power of the Hindu right-wing in Indian politics, Ahmad (1997a) occupies a Marxist and secularist position, which means that he rejects nationalistic views of culture and emphasizes universal principles. By contrast Esteva and Prakash (1998), also writing of India, advocate the opposition to global capital through local initiatives by popular movements. They maintain that there *are no* universal values, principles or morals. The theoretical position behind the two views is radically different, as is the action that would be taken in the light of these positions. This sort of clash is obviously something that cultural geographers need to address, even if they cannot pretend to provide satisfactory answers.

> [F]or those proselytizing the 'secular morality' of human rights, it is not considered dogmatic to ask of anyone, regardless of religion, race, colour or culture: 'Don't you believe in human rights (the modern "sacred" cow) for all men, children and women?' They fail to see that their faith is as threatening to all the diverse cultures of the world as the Trojan horse was for the people of Troy.
>
> We come from cultures and traditions for which the concept of human rights is not only alien, but, furthermore, actually incommensurable with the central cultural ideals or virtues. Consequently, we find this modern (secular?) faith to be as dangerous to our cultures as those that sought to convince us that our peoples were pagans; that we did not have the right Gods or the One True God; that we were strange, and primitive and uncouth for praying to monkeys, elephants or the phallus of Shiv. Resisting the monoculture of any one true global God or religion, we celebrate grassroots groups that do not fall victims to this Trojan horse of recolonization.
>
> There is, we realize, enormous violence, abuse and suffering everywhere CNN turns its global 'eyes' – from Bosnia and Berlin to Beijing. Even as we mourn this tragic reality, hoping for its amelioration, we recognize that our own cultures are neither superior nor unique in possessing moral concepts for correcting our inhumanities; including, for example, India's 'dowry deaths', which receive international attention and attack from human rights activists. Intent on exporting 'human rights' to Hindus, these activists fail to take note of *aadar, sammaan, shradha, izzat, hak, dharma, ahimsa*, among other moral concepts that enjoin human decency as well as condemn violence against women and animals. Many of these words are not easily translated; they lack English equivalents. They offer important clues about Hindu morality, its *dharma* and virtue . . .
>
> Yet, Hindu *dharma* (for appropriate and respectful treatment of women and others) did not receive the attention and importance given to the concept of 'women's rights' in Beijing. Why?
>
> (Esteva and Prakash 1998 pp. 118–19)

One needs to try to answer why a theoretical position based upon a concept such as *dharma* seems not to be universalizable (and to wonder whether there is not a strange contradiction within Esteva and Prakash's argument).

TASK

> • Do you agree with Esteva and Prakash (1998) that virtue and morality are local?
> • What are the political implications of assuming a radically localist theoretical perspective?

INDIVIDUAL PERSPECTIVES

This short chapter simply wants to raise questions as to why people explain things the way that they do and to argue that a major part of what is known as culture is concerned with assembling the values, communicative devices, actions and aesthetics which fit in with favoured explanations.

TASK

> Given below are various accounts of urban life in the West. All recognize problems in urban living.
>
> • Try to decide what the main issue is for each writer and what, by implication, he/she thinks should be done about it.
> • Identify the similarities and differences between the ways of describing and explaining offered by the different authors.

In 1989, a private organization that manages Central Park, the Central Park Conservancy, demanded demolition of the Naumberg Bandshell, site of popular concerts from the 1930s to the 1950s, where homeless people gathered. Similarly, the Bryant Park Restoration Corporation started cleaning up the midtown business district by adopting the social design principles developed by Whyte. Whyte's basic idea is that public spaces are made safe by attracting lots of 'normal' users. The more normal users there are, the less space there will be for vagrants and criminals to manoeuvre. The Bryant Park Restoration Corporation intended their work to set a prototype for urban public space. They completely reorganized the landscape design of the park, opening it up to women, who tended to avoid the park even during daylight, and selling certain kinds of buffet food. They established a model of pacification by cappuccino.

Central Park, Bryant Park, and the Hudson River Park show how public spaces are becoming progressively less public: they are, in certain ways, more exclusive than at any time in the past 100 years. Each of these areas is governed, and largely or entirely financed, by a private organization, often working as a quasi-public authority.

(Zukin 1995 p. 28)

The perverse rationality of real estate capitalism means that building owners garner a double reward for milking properties and destroying buildings. First they pocket the money that should have gone to repairs and upkeep; second, having destroyed the building and established a rent gap, they have produced for themselves the conditions and opportunity for a whole round of capital reinvestment. Having produced a scarcity of capital in the name of profit they now flood the neighbourhood for the same purpose, portraying themselves all along as civic-minded heroes, taking a risk where no one else would venture, builders of a new city worthy of the populace . . .

The economic geography of gentrification is not random; developers do not just plunge into the heart of slum opportunity, but tend to take it piece by piece. Rugged pioneersmanship is tempered by financial caution. Developers have a vivid block-by-block sense of where the frontier lies.

(Smith 1996 p. 23)

However one interprets this globalization process, its impact on the urban fabric and on urban life has been enormous and it is likely to continue to be so in the future. Many cities were world cities before the 1960s, but since then, if you will excuse the turn of phrase, the globalization of the city has itself become increasingly globalized. Every contemporary city is to some significant degree also a world city in much the same way as it is postmodern. Everywhere the local is becoming globalized and the global is becoming localized, giving birth to another of those neologisms that attempt to describe the distinctive results of contemporary restructuring processes, the concept of 'glocalization'.

The *third geography* has generated perhaps the greatest flurry of neologisms to describe its newness. A short list would include such terms as megacities, outer cities, edge cities, metroplex, technoburbs, postsuburbia, technopolis, heteropolis, exopolis. What is being described in these terms is a radical restructuring of urban form and of the conventional language we have been using to describe cities. Following on the odd combinations of opposing tendencies that have been used to describe the first two geographies of restructuring – deindustrialization and reindustrialization, globalizing the local and localizing the global – the third restructuring can be described as a combination of decentralization and recentralization, the peripheralization of the center and the centralization of the periphery, the city simultaneously being turned inside out and outside in.

(Soja 1995 p. 131)

The mere reiteration of the phrase 'cities as if people matter' will not get us very far. The humanistic perspective is not in itself a solution or an end-point of explanations, but a starting position for those seeking sustainable, liveable solutions to our social malaise. And there is malaise, whether it be experienced in communal revolt, random acts of violence, anonymous bouts of depression or agonizing loneliness. Some of our ills are not new. People have always been depressed, and street violence is not simply a product of the twentieth century. Where we are in time is not so much the problem – more a source of emancipation. We now have more resources, knowledge and abilities at our collective disposal than ever before. It seems such a waste not to use the opportunities presented to us by our position in history and our location in society to build better societies and to create more humane cities.

(Short 1989 p. 4)

Successful control presumes a power to exclude unwanted elements. Fine-tuned ethnic, religious, racial, and status discriminations are frequently called into play within such a process of community construction. Furthermore, political organization takes a special form, generally expressive of a culture of political resistance and hostility to normal channels of political incorporation. The state is largely experienced as an agency of repressive control (in police, education, etc.) rather than as an agency that can be controlled by and bring benefits to them . . . Political organizations of a participatory sort are . . . weakly developed and politics of the bourgeois sort [is] understood as irrelevant to the procuring of the use values necessary for daily survival. Nevertheless, the state intervenes in such communities since they are vital preserves of the reserve army of the unemployed, spaces of such deprivation that all sorts of contagious social ills (from prostitution to tuberculosis) can flourish, and spaces that appear dangerous precisely because they lie outside of the normal processes of social incorporation.

Contrast this with the practices of affluent groups, who can command space through spatial mobility and ownership of the basic means of reproduction (houses, cars, etc.). Already blessed with abundant exchange values with which to sustain life, they are in no way dependent upon community-provided use values for survival. The construction of community is then mainly geared to the preservation or enhancement of exchange values. Use values relate to matters of accessibility, taste, tone, aesthetic appreciation, and the symbolic and cultural capital that goes with possession of a certain kind of 'valued' built environment.

(Harvey 1989a p. 266)

TASK

- If you did these readings as a class activity, discuss your interpretations with other people; it is quite likely you will find that your opinions do not exactly converge.
- If this is the case, try to decide what it is about you as individuals which makes you construct different interpretations.

Some people will maintain that they think the way they do because of their religious faith; others will maintain that their ideas have been shaped by their ethnic background, their gender, their class position; some will look to their parents or their schooling as the source of their values; some will refer to an event which transformed their perspective; others will hold the belief that all individuals are absolutely unique and they explain things the way they do 'because I'm me'; some people are fatalists who explain by trying not to explain. Theoretical positions will be chosen which are most compatible with one's normal way of seeing the world.

The one thing we cannot do with the realization that there are conflicting viewpoints is to decide to hold all of them at once, even though we may move between them over time. People who favour compromise over conflict have to realize that a compromise position is itself a position. A major reason why we need to grapple with theory is to try to sort out the conflicts and contradictions in our own arguments, and to change our positions if necessary. Another reason is that, though I have been stating things very baldly, people hold theoretical positions which are subtle and nuanced and few writers announce them loudly. In order not to be persuaded by people with different ideologies from oneself one needs to be sensitive to ideology.

GLOSSARY

Belief   Something one assumes as an article of faith to be true (but cannot prove).

Common sense   Locally unquestioned way of explaining.

Concept   An abstraction which helps one to understand.

Ideology   Belief.

Knowledge   That which is held to be true.

Perspective   Point from which one looks.

Phenomenon   Something which can be experienced through the senses (plural 'phenomena').

Reification   Act of treating an abstraction as if it were a thing (verb 'reify')

Theory   Knowledge which is unverified (tends to mean way of explaining).

FURTHER READING

Jenks, C. (ed.) (1998) *Core Sociological Dichotomies*. London: Sage.

Larrain, J. (1994) *Ideology and Cultural Identity: Modernity and the Third World Presence*. Cambridge: Polity.

Squires, J. (ed.) (1993) *Principled Positions: Postmodernism and the Rediscovery of Value*. London: Lawrence and Wishart.

3 Humanistic and behavioural geography

Carol Ekinsmyth and Pamela Shurmer-Smith

Though contemporary cultural geography owes much to the behavioural and humanistic approaches which constitute the previous generation of the field, very few researchers are adopting either of these approaches today. The main reason for the rejection has been that they have often implicitly drawn upon superorganic notions of culture and essentialist ideas about the individual, which have become less intellectually acceptable in the light of more recent considerations of the human subject. Whilst differentiating between humanistic geography, with its roots in phenomenology, and behavioural geography, grounded in psychology, Steve Pile draws attention to the similarities between them in that, 'They share a commitment to place "man" at the centre of "his" world' (1996 p. 16). It is arguable that, though elements of these approaches can be used as a stage to pass through when trying to understand the development of radical and post-structuralist positions, undergraduate research projects from a behaviouralist or humanistic perspective can rarely engage with the cutting edge of cultural geography. A major difficulty with putting 'man' at the centre of 'his' world (even in inverted commas) is that by no means all people prioritize individuals over groups. Individualism is a construct which can be associated with modernism and the Western middle-class values it engendered; it is most ideally formulated in the independent, masculine, hero.

The structuralist anthropologist Claude Lévi-Strauss is reputed to have sneered that phenomenology was a 'shop-girl philosophy', because he thought that saying 'It all depends on your point of view' was a bit too easy. However, many people, especially those with strong religious convictions, do hold a fundamental belief that all human beings are essentially unique individuals who see the world in their own special way. Humanistic geography can seem an appealing way of incorporating such beliefs into academic work; the trouble is it all too often becomes nothing more than description of particular actions in particular places, combined with a celebration of the human spirit. We argue that this may be appropriate as a starting place but not the end point of research, but we are willing to concede that there are plenty of practising academics who would disagree.

The heyday of humanistic geography was in the 1970s and it is usually seen as a reaction against the emergence of geography as a *spatial science* and the accompanying reliance on quantitative research methods. It was, however, also a reaction against the growing popularity of Marxist approaches in human geography, as is evidenced in the highly influential collection edited by Ley and Samuels (1978). Both spatial science and Marxist geography could be regarded as taking individuals out of the frame, as they emphasized collectivities, trends, masses and structures. Humanistic geography, by contrast, placed its emphasis upon personal experience and understanding of landscape and environment, most clearly expressed in its concept of 'sense of place'.

In theoretical terms, as Cloke et al. (1991) show, humanistic geography drew upon an eclectic range of sources in philosophy. Existentialism (with its insistence that simple human existence precedes society and culture and its assumption that humans are compelled every day to choose the nature of their being, even if the choice is only to do nothing) and phenomenology (which concentrates on the ways in which the world appears through the senses) are usually seen as the major inspirations for the view that embodied human beings should be the primary concern of human geographers. Humanistic geography, then, has as its starting point the assumption that there is no possibility of studying a 'real' world outside human feeling, thought and understanding. The (impossible) task of the humanistic geographer was to access this interior experience. Behaviouralist psychology seemed to offer a route to this interior, but it required one to assume that the individual was a product of cognitive processes based on response to external stimuli which could, literally, be charted. So Gould and White's (1974) notion of mental maps purported to grasp the ways in which people perceived familiar and unfamiliar environments by asking standardized questions and by asking people to draw sketch maps.

Projective techniques: word association tests (where individuals say what comes into their mind when they hear a given place name, thereby indicating what the place means to them); construction techniques like the Thematic Apperception Test (where individuals are given a picture of a particular environment and required to construct a story about what is happening in the picture); cloze procedures (where individuals fill in the missing parts of a stimulus, for example the missing parts of a map, thereby giving an indication of the completeness of their overall image).

(Walmesley 1988 p. 47)

BEHAVIOURAL GEOGRAPHY

Behavioural geography, at its height between the mid 1970s and the mid 1980s, represents a period when geographers became excited about psychology, especially concepts such as 'cognition', 'perception', 'development' and 'learning', and their effect on spatial behaviour. The seeds of the subfield were sown in the 1940s, 1950s and 1960s. Its early proponents expressed a concern about the normative models of human behaviour that were influential at the time, such as those of Losch, Weber and Von Thunen. Behavioural geographers were centrally concerned with human decision-making and resultant spatial behaviour, and they recognized that 'there are many situations in geography in which these descriptive devices or idealizations [those of normative models] are clearly inappropriate and there is no alternative but to incorporate very specific statements about the cognitive processes involved in the act of decisions' (Harvey 1969 p. 35).

To be able to 'incorporate very specific statements' about 'cognitive processes', such processes had to be studied. Thus geographers began to explore what Wright (1947) described as the *terrae incognitae* in the minds of women and men. Kirk (1952; 1963) introduced into geography the concept of the 'behavioural environment' as distinct from the objective or phenomenal environment, and thus provided behavioural problems with a geographical context. He was also a pioneer amongst geographers in attempting to introduce the principles and concerns of psychology into the geographical arena. The concept of the 'behavioural environment' was derived from a psychologist, Kurt Lewin, who was one of the first people to talk about 'perceived space' (he called it 'lifespace'), 'real space', and the gap between them. He considered that as learning proceeded, a person's space got closer and closer to 'real space' (Lewin 1936).

Most effort in the early years of the development of behavioural geography was invested in attempting to understand the outcome of 'cognitive processes', rather than in trying to appreciate the nature of the actual processes themselves (this became a focus later on). This outcome was a person's 'behavioural environment', which consisted of their cognitive representation of space (the 'world in the head', the 'map in the mind'). Various concepts were used to denote these cognitive representations, but the one most often employed was the 'image'. This concept gained acceptance in geographical circles after the work of Boulding (1956) who suggested that individuals over time build up mental impressions of the world (images) which act as the basis for their behaviour in space. These images were not simply derived from physical interaction with the phenomenal environment but were much richer impressions, gained over the years from an individual's physical and social milieu. The images that constituted a person's behavioural environment were to be the explanatory variables in the modelling of spatial decision-making. This undermined the previous indispensability of the normative assumptions of 'rational economic man', and decisions based on perfect knowledge in isotropic environments. As other chapters in this book show, critics of positivism today argue that this effort did not go (anywhere near) far enough to undermine normative assumptions and assertions.

From the outset, behavioural geography was policy oriented. This is reflected in the early concentration of research effort on natural hazards and people's adjustment to them. By the mid 1980s, research efforts in the sub-field were diverse in nature. This led Goodey and Gold to speak of the 'blurred and uncertain identity of this field of study' (1985 p. 587). The relationship between behavioural geography and humanistic geography also remained unclear at the time. Some authors saw humanistic geography as an approach within behavioural geography. They typified behavioural geography as having a schism between those adhering to positivist approaches and those preferring humanist approaches (Walmsley and Lewis 1984; Goodey and Gold 1985). Today, it seems more sensible to consider behavioural geography as the positivist approach to human behaviour (relying as it did upon attempts at systematic observation and measurement), as distinct and separate from humanistic geography which had, at times, similar goals but a different focus. It certainly had very different methods.

The outcome of effort in behavioural geography was a body of work that attempted to understand the reasons for the environmental behaviour of people. Investigations into perceptions of flood hazard, for example, would help to explain why people change address, or experience stress in their homes. Investigations of perceptions of holiday destinations might help to explain holiday destination choice. Understanding people's misunderstandings of distance and direction might help to explain why some do not use the facilities, such as shopping centres, which are closest to them in real distance. This could help planners in their decisions about where to locate facilities. Having knowledge of people's 'mental maps' (which contain environmental perceptions – of threat for instance), and understanding how these mental maps are acquired, might aid understanding of how to create environments that are easy to 'read', that feel safe, that are more pleasant. These were some of the beliefs and goals of behavioural geographers at the time.

By and large, behavioural geography has disappeared (although there are some people in North America whose work has similar goals). It never really lived up to its promises, and indeed, many would argue that it never could have done. Its underlying aim, to help the understanding of the link between the subjective personal world of the individual and his or her spatial behaviour, was never really accomplished. Behavioural geographers would claim that their work did accomplish a deeper understanding of the 'world in our heads' (although even this is open to debate), but this was only a partial understanding, and it failed to make the link between 'image' and subsequent behaviour. In fact, it was easier to make the link in reverse, that is, how behaviour affects the 'world in our heads'. Without this link, critics could ask of the claims of behavioural geographers – 'so what?' But even more fundamentally, the sub-field, based as it was around positivist philosophy and normative assumptions, could not survive the ever more persuasive and sophisticated critiques of positivism that were being increasingly aired from the mid 1980s on. Of course, many behavioural geographers had been attracted to the field in the first place because of a dissatisfaction with the sterile, 'human-less' nature of the positivist geography of the 1960s and early 1970s, so they were in a prime position to become their own critics. Many behavioural geographers have maintained their interest in the subjective personal world, but are now studying this in different ways and for different reasons.

> If images are important we should find some way of measuring them. Even a rough measure of a significant variable is better than a precise measure of something that does not matter.
>
> (Saarinen 1979 p. 465)

> Probably the most basic problem in perception research concerns the lack of any acceptable standards or criteria against which such elusive mental phenomena as environmental images and preferences can be evaluated and checked. Lacking any obvious or direct verification methods, some geographers have fallen into the trap of adopting the notion that anything that can be measured in one way or another must be real and accurate. Little consideration has been given to the philosophical and methodological tenets that underlie the issue of measuring mental images. Indeed, no consensus exists among researchers about the terms of their measurement techniques and no concentrated attempt has been made to develop a standardized methodology.
>
> (Bunting and Guelke 1979 p. 454)

> With systematic observation . . . the goal is for properly trained observers to produce identical protocols, given that they observed the same stream of behaviour.
>
> (Bakeman and Goffman 1997)

TASK

- How do the authors of these accounts differ in opinion?
- What are the underlying assumptions of the authors about possible methods for exploring the mental images of people? (Consider the words used such as 'measurement', 'variable', 'verification'.)
- Do you think Bakeman and Goffman's goals are *really* achievable? (Should any doubt stop us from trying?)

TASK

- Make a list of the reasons why attempts to understand the nature and content of 'mental maps' might be difficult.
- Now list reasons why behavioural geographers found it difficult to make the link between knowledge, perception and subsequent behaviour.
- Do you think that research using conventional scientific methods is capable of gaining an understanding of the 'subjective personal worlds' of individuals?

SUBJECTIVITY

As the concerns relating to behavioural geography built up, there was a turn towards encouraging geographers to dig deep into their own perceptions and preferences relating to the meaning of landscape (Meinig 1979; Relph 1976; Tuan 1976) which encouraged an authorial subjectivity of a kind that might be considered arrogant, but which people starting from similar social and intellectual positions seemed to welcome.

Placelessness describes both an environment without significant places and the underlying attitude which does not recognise significance in places. It reaches back into the deepest levels of place, cutting roots, eroding symbols, replacing diversity with uniformity and experiential order with conceptual order. At its most profound it consists of a pervasive and perhaps irreversible alienation from places as the homes of men: 'He who has no home now will not build one anymore,' Rilke declared.

(Relph 1976 p. 143)

'Open' and 'closed' are spatial categories meaningful to many people. Agoraphobia and claustrophobia describe pathological states, but open and closed spaces can also stimulate topophilic feelings. Open space signifies freedom, the promise of adventure, light, the public realm, formal and unchanging beauty; enclosed space signifies the cozy security of the womb, privacy, darkness, biologic life. It is tempting to speculate on the relation of these things to certain profound human experiences considered phylogenetically and ontogenetically. As a species, man's primate ancestors migrated out of the womb-like shelter of the tropical forest to the more open and unpredictable environment of the open savanna. Individually, every birth seems to move out of the dark protective womb to a bright world that seems far less accommodating.

(Tuan 1976 pp. 27–8)

TASK

- Are you convinced by Relph and Tuan's interpretations?
- What sorts of assumptions underlie their ability to distinguish good and bad spaces and places?
- Do you think that mothers are likely to employ womb metaphors the same way as people who have only experienced wombs as sons?
- Think about the other metaphors they use and wonder how generalizable they are.

One of the main problems with this shift to subjective writing was precisely the phenomenological issue of *point of view* (this needs to be thought both metaphorically and literally). Much of what was produced focused unambiguously on the *man* in humanism, in particular the educated, white, middle-class man; the *man* seeking prospect and refuge (Appleton 1975), the *man* establishing a dwelling (Seamon 1985). The issue of the constructedness of *men* as political entities and the occupants of structural positions rarely emerges.

> Rest anchors the present and future in the past and maintains an experiential and historical continuity. From the vantage point of human experience, the deepest manifestation of rest is *dwelling*, which involves a lifestyle of regularity, repetition and cyclicity all grounded in an atmosphere of care and concern for places, things and people . . . Dwelling can be seen as an aim to strive for.
>
> (Seamon 1985 p. 227)

TASK

- Can you read any political ideology into Seamon's statement?
- What world view is expressed in his 'concern for places, things and people'?
- Would you see this restful state as universally attractive?

Until issues of authority were tackled with the textual awareness that came through deconstruction, emphasis on a sensitive reading of landscape or representation of landscape would inevitably privilege certain interpretations over others. Quite often, one finds other people's meanings to be meaningless and it is an interesting exercise to try to think how some of Relph's 'authentic' places might seem for various categories of 'outsiders'. Though 'sense of place' is an enticing concept, it ought to be obvious that there is no single genuine essence buried in any place to be mined by the trained cultural geographer.

One day Katy Bennett and I were walking together down a crowded Delhi street. It was only when she suddenly announced that she couldn't stand any more, that I realized that Katy had been suffering silently as lecherous hands reached out to her, whilst I had obliviously been enjoying being immersed in the vibrant throng. What a difference a 25-year age gap can make to a woman's sense of place.

(Pamela Shurmer-Smith)

TASK

Think of a place of social activity where you feel uneasy and examine your reasons for your discomfort. Think about the people who regard this place as congenial and why. Then try it the other way about. Is it personal or is it to do with your social attributes?

Humanistic geography was useful for alerting people to the sensory and personal and for making readers aware that there are multiple meanings. However, in its emphasis on 'individuality' it closed its eyes to the politics behind the different ways of understanding place and space: a space one kind of person feels able to dominate is likely to be one where another type of person feels dominated; a place where one category of person feels safe is where another may feel excluded.

As later chapters on ethnographic methods and textual analysis will consider, the collection and interpretation of the plurality of meanings of places and spaces can help one grasp the way in which multistranded local knowleges are constructed. These many knowledges constitute a local body of meaning not just because they are randomly assembled in the same place, but because they are forged in the light of one another. Different meanings and values can be complementary and complicit, but they are often oppositional, and there are also those which just make the best of what is left over from stronger forces' struggles. Simply to describe the various human senses of place is to leave the important part of the cultural work undone; one's task must be to show how they ravel together, support and undermine one another and constitute a dense fabric.

GLOSSARY

Behavioural geography   Assumption that human spatial behaviour could be explained by using psychology.

Ethnography   Literally writings about people, but has come to mean research based on personal contact with people.

Humanistic geography   Assumption that unique individuals influence (and uniquely experience) place.

Phenomenology   Philosophy based on recognition of different viewpoints.

Sense of place   The unique character of a particular place.

Text   Any representation.

FURTHER READING

Bunting, T. and Guelke, L. (1979) 'Behavioural and perceptual geography: a critical appraisal', *Annals of the Association of American Geographers*, 69: 448–62.

Ley, D. and Samuels, M. (eds) (1978) *Humanistic Geography: Prospects and Progress*. Chicago: Maarouffa.

Relph, E. (1976) *Place and Placelessness*. London: Pion.

Schutz, A. (1967) *The Phenomenology of the Social World*. London: Heinemann.

Seamon, D. and Mugerauer, R. (eds) *Dwelling, Place and Environment: Towards a Phenomenology of Person and World*. Dordrecht: Martinus Nijhoff.

Walmsley, J. and Lewis, G. (1984) *Human Geography: Behavioural Approaches*. Harlow: Longman.

4 Marx and after

Pamela Shurmer-Smith

Marxist thought occupies a very important position in contemporary cultural geography, even though the number of people who would unhesitatingly label themselves 'Marxist' is dwindling. The realization that much could usefully be borrowed from cultural studies, with its heavy reliance on the Gramscian notion of cultural hegemony, marked the break from the old superorganicist approach.

The 'new' cultural geography recognizes that culture is always political. It is not a matter of dragging politics into culture; it is rather that meanings, values and cultural practices are always and already political. Don Mitchell (2000) makes this crystal clear in his critical introduction to the subject.

> It is no accident that the culture wars being fought *now*, over the end of the twentieth and beginning of the twenty-first century, have a particular shape and style to them; they are battles over cultural identities – and the power to shape, determine and, literally, *emplace* those identities . . . Any culture war – now or in the past – is both a reflection of, and an ongoing contribution to, the geographies we build in the world. Each successive battle transforms the geography in which it takes place and therefore creates new contexts – new geographical situations – within which the next round of struggle occurs. But these geographies are not only – and can never be only – 'cultural'. Instead they are part of a recursive set of relationships between what we call 'culture' and the changing political and economic fortunes of different places and different peoples.
>
> (Mitchell 2000 p. 11)

At the heart of Marxist thinking about culture is the assumption that the moral and aesthetic values of a society are configured in such a way that they reinforce its economic and political structures. It is no coincidence that the word 'value' operates in both the economic and moral realm; that European languages use a word meaning 'expensive' as a term of en*dear*-ment and that 'cheap' does not always refer to price. Godelier (1986) argues that the relations of production will influence which aspect of society is seen as most important: so, for example, if families are the groups responsible for producing wealth, kinship will be revered; if a priesthood controls wealth, then religion will move to centre stage.

Marx alerted his readers to the role of religion in maintaining the values which reinforced the position of the wealthy and powerful, but anthropologists such as Lanternari (1963), Worsley (1970) and Comaroff (1985) have shown how religion can also be used in precisely the opposite way, to mobilize the ideas and action which will bring about change. But we can see secular systems of ideas and modes of communication as also performing the same functions of both reinforcement and resistance.

COMMODITY FETISHISM

Arguably, Marx's main contribution to thinking about culture is his concept of *commodity fetishism* – the realization that things artificially hyped in a market context become objects of desire as a consequence of their economic value (rather than the other way about, which would seem to be more sensible). In the concept of commodity fetishism we can see Marx laying down the basics of the idea of consumer culture, whereby people go on striving for more things long after they have satisfied their needs. The section on Baudrillard in the next chapter will consider the way in which, in post-industrial societies, this takes off into what he calls 'viral' value where there does not even have to be any tangible object at all.

> A commodity appears, at first sight, a very trivial thing and easily understood. Its analysis shows that it is, in reality, a very queer thing, abounding in metaphysical subtleties and theological niceties. So far as it is a value in use, there is nothing mysterious about it, whether we consider from the point of view that, by its properties, it is capable of satisfying human wants, or from the point that those properties are the product of human labour . . . a table continues to be that common, everyday thing, wood. But so soon as it steps forth as a commodity, it is changed into something transcendent. It not only stands with its feet on the ground, but in relation to other commodities, it stands on its head, and evolves, out of its wooden brain a grotesque idea, far more wonderful than 'table-turning' ever was.
>
> (Marx 1970 p. 76)

As Marx indicates, commodification is not just a matter of paying for things which once were free; it changes the value in all senses. Commodification aims to stimulate desire for things and experiences which goes beyond need and want. The extract from Giulianotti (1999) focuses on the way in which football has become different since it became much more expensive to watch and players started to be paid very large sums of money. The new expensiveness has gone hand in hand with a glamorization of the sport and the development of an international star system which goes well beyond mere respect for skill.

Since 1990, the structural nexus of football and the working classes has been strongly undermined. Football clubs and the police are less tolerant of expressive forms of support. Ground redevelopment has replaced the old terraces with more expensive, family-friendly stands. Those locked out must forfeit a hefty subscription fee to watch on television. Merchandising and share issues mean that clubs pursue wealthier, national fan groups rather than satisfy local supporters. On the park, the local one-club heroes have become peripatetic national or international 'celebrities', drawn increasingly from affluent suburbs rather than poor housing estates.

Concomitantly, the working class itself has undergone major structural changes since the 1970s. Deindustrialization and the rise of the service-sector economy have reduced the industrial working class and expanded the white-collar workforce. The structural boundaries between the old lower middle classes and the affluent upper working classes have become very blurred. A dispossessed underclass is sedimented at the base of the new class hierarchy. In the post-modern era, this underclass and the class strata just above are most visibly excluded from football's brave new world. UK football's new target audiences include family groups, the first generation middle classes and the young metropolitan elite. These developments have certainly enlivened the cultural politics of football. The new middle classes have contributed particularly to the UK game's fashionability throughout the 1990s.

(Giulianotti 1999 p. 51)

A Marxist cultural geographer might like to explore the implications of the commodification of sport, whether locally, nationally or globally and the way in which this plays into the commodification and promotion of sporting venues. I wonder whether the equivalent of the Ali–Foreman 'Rumble in the Jungle' (1974 in Zaire) could take place today, with all the implications for loss of local economic spin-off (Marqusee 1999).

Even the *act* of shopping can be commodified when going shopping becomes an experience beyond mere provisioning. Studies of mall culture amply demonstrate that the mall (and before it the department store and the arcade) aims to stimulate the desire for commodities by setting them in a special, culturally legible, context (Shields 1991; Holbrook and Jackson,

1996b). The commodified place is also intentionally exclusionary of 'unde-sirable' people. There is an aesthetics, a politics, a sociology and a morality attached to commodification; shifting something into the realm of com-modities has deep cultural implications.

CULTURAL HEGEMONY

Gramsci made Marx's theories of culture relevant to the mass societies of the twentieth century. He wrote from prison during the years of fascist rule in Italy and his underlying purpose was to understand how decent, ordinary people had come to believe in such an abhorrent system, so that he could mobilize opposition to it. Gramsci's (1971) great contribution to the study of culture was to argue that, although practices and values accorded with the maintenance of existing systems of wealth and power, people were not coerced into these; they were, instead, persuaded. This was done by means of systems of education and the media establishing a viewpoint which became accepted as 'common sense'. He coined the term *cultural hegemony* to describe the situation in which people subscribe to knowledge systems which are actually to the advantage of people with superior power and wealth, rather than striving for their own interests. It is in the nature of cultural hegemony that one does not notice the pro-cess of value formation. If one were to be conscious of how values are shaped, it would not be an effective device. Thus, for example, most people think that the newspaper they choose to read is fairly objective, whereas they are likely to see others as being left- or right-wing propaganda rags. For there to be cultural hegemony, the message and the medium of its communication have to be acceptable, framed within the existing values and experiences of those receiving it; they have to seem reasonable. Hegemony, therefore, becomes a dialogue through which the various kinds of 'common sense' are negotiated to the advantage of the prevailing power structures.

TASK

> - Swop your usual Sunday newspaper with someone who has a radically different political position from yourself.
> - Mark all the features you would regard as having a cultural geography dimension (access to the countryside, urban lifestyles, immigration etc.).
> - Note the tone of the articles, the terminology used and the examples offered.
> - When you disagree with the interpretations or are angered by the style, decide why this is.
> - See what the other person thinks of your views.

It is not only in news reporting that we can see hegemonic processes at work. Every mediated communication has a political message; every novel, film and TV soap structures its knowledge in ways which reinforce or challenge existing order, moral systems, economic and political relationships. Sometimes this is very subtle, sometimes it is blatant, but as with the news we tend to be aware of the process only when it opposes our own ideological position. The distance of time can often help make hegemonic processes more clearly observable.

TASK

> - View a mainstream British film made in the 1930s or 1940s.
> - Observe how the social class of characters is signified.
> - Classify the characters (e.g. heroic, romantic, villainous, supporting, comic) and think about the class positions assigned to them.
> - Decide whether you think that the film communicates anything about what were regarded as 'normal' assumptions about status.
> - Try the same exercise with an American film of the same period and then a British film made in the last two years.

EXCLUSION

The cultural geography influenced by Marxism concentrates on the implicit inequities in the ways in which places are constructed and spaces used. The market is obviously important in excluding some people from expensive places and things, but beyond this we can see other kinds of exclusion based on ideas about which sorts of people are regarded as entitled to which sorts of spaces. The question of urban gentrification and upper-middle-class counter-urbanization can be seen as clear examples of ways in which the cultural preferences of affluent and educated people lead to appropriation of spaces which were formerly occupied by poorer people, who are then marginalized. Not just price itself or improved material conditions, but also ideas about exclusiveness, are used to change the attractiveness of places. Smith (1996) gives an excellent account of the interplay of economics, politics and cultural forces in the colonization of urban spaces. Terms like 'redevelopment', 'renewal' and 'improvement' are themselves cultural devices which make those who resist change seem out of step.

Marxist cultural geography is sensitive to issues of social class as it relates to space and place, but it is clear that in post-industrial societies old class formations can become indistinct and new kinds of advantage and deprivation emerge. The concept of consumption cleavage (Saunders 1984) has usually been associated with the politics of the right, but it valuably alerts one not only to the ways in which there is an element of investment in certain consumption strategies, but also to the significance of the composition of households in influencing standards of living. If one thinks about consumption cleavages alongside Bourdieu's (1984) concept of *cultural capital*, including the investment in knowledge and taste, it becomes clear that today's Marxists need to think about more than simple class positions based upon relations of production when they are explaining exclusions and deprivations.

Geographers have paid relatively little attention to the work of André Gorz, a Marxist social critic who theorizes post-industrial society. Gorz (1982) foresaw the demise of the working (and therefore any other) class and suggested that unless there were permanently to be a high proportion of unemployed and excluded people, there needed to be a new mode of social equality based on the sharing of scarce work. In *Ecology as Politics* (1980) Gorz provided a Marxist cultural formulation of the intimate relationship between the human and ecological costs of overdevelopment. Building on the idea of commodity fetishism, he advocates moving away from the market economy which results in unnecessary production. It is obvious that such a radical shift in ideas relating to work and environment requires considerable change in the way in which people think about each other and the world; the shift would be as cultural as it is economic and political. This is certainly a political philosophy which geographers, regardless of their own stance, need to have an opinion about.

AFTER MARX

We can see Marxist concerns with inequality of wealth, power and opportunity as extending beyond simple formulations of class into cultural politics based on such things as gender and sexuality, race and ethnicity. They have implications for various minority rights groups such as disabled people, aged people and children, people advocating alternative lifestyles, again, not class based categories. Laclau and Mouffe (1985; Laclau 1994) have been at the forefront of the post-Marxist movement which emphasizes minority rights and claims that there is a need for a democratic revolution which sees a movement away from what Indian theorists such as Ahmad (1996) and Vanaik (1997) call 'majoritarianism'. Such ideas have also been propounded by anti-development theorists who put their faith in local solutions.

Discrimination and prejudice are inherently cultural but their outcomes are experienced in citizenship, labour markets, housing, personal relationships, voting patterns and a host of other important aspects of life. The consciousness of minorities (including the ways in which people are constituted as minorities) has led not only to the study of difference but also to a recent interest in geographies of exclusion (Sibley 1995; Cloke and Little 1997).

> If we think about the question of residential segregation, which is one of the most widely investigated issues in urban geography, it could be argued that the resistance to a different sort of person moving into a neighbourhood stems from feelings of anxiety, nervousness or fear. Who is felt to belong and not to belong contributes in an important way to the shaping of social space. It is often the case that this kind of hostility to others is articulated as a concern about property values, but certain kinds of difference, as they are culturally constructed, trigger anxieties and a wish on the part of those who feel threatened to distance themselves from others. This may, of course, have economic consequences.
>
> (Sibley 1995 p. 3)

TASK

- Think about the meaning of the word 'exclusive' as a positive term when describing lifestyle.
- Now try substituting the word 'exclusionary'.
- Does this substitution clarify the connection between social, cultural and economic value?

Although post-Marxists place considerable emphasis on the power of local social movements to effect change and to forge a new non-party politics, it is undeniable that apathy is a far stronger force accompanied by low-key resentment as the main form of resistance. One needs to return to Gramsci to think why this should be the case.

TASK

- Search your local newspaper and fly-posters for public meetings on local issues of concern to geographers (neighbourhood protection meetings, protests about a new development, campaigns for an amenity, anti-pollution groups etc.).
- Attend one of these.
- Note what sorts of place the meetings take place in and the numbers of people who attend.
- Can you generalize about the kinds of people who attend?
- Is there any conflict? If so, how is it expressed?
- What sort of language is used – official, eloquent, colloquial?
- How do you explain the event in terms of your perceptions of local culture?

POLITICAL CULTURE

Strikes, boycotts, civil disobedience and armed struggle are the most obvious forms of resistance to unpopular political and economic regimes, but every resistance movement knows that it must also concentrate on the cultural front to establish its counter-hegemony. To be successful, resistance has to become normal and moral; the ideas have to be propagated (hence propaganda). There was never a revolution without its own artists, without its new morality and without its special ways of communicating.

Access to a printing press, or just a duplicator, is always seen as important to a resistance movement, but graffiti and posters are also effective ways of propagating a message. Hectoring propaganda is effective only in sustaining the enthusiasm of those who already subscribe to the goals of the movement, otherwise the usual techniques of cultural hegemony apply. Fredrik Jameson (1995) refers to the way in which even apparently mainstream film can convey messages of resistance and this would apply to other art forms. The recited poetry of Pablo Neruda (1978), for example, was a very real rallying device for the poor and illiterate miners of southern Chile who committed it to memory. Although much of this poetry is concerned with working conditions, his popularity came from the way in which he could capture and dignify the sentiments of the people when he wrote about love or nature.

The Internet has gained considerable attention as a tool for messages of resistance. It is a very effective mode of communication, but only as a means whereby movements get in touch with sympathizers; it is unlikely that it can be used to convert people. Its great advantage is that information can be up-to-date and there is no need for expensive mailing. One of the other advantages of the Internet is that small and impoverished groups can link into bigger sites. Many groups in isolated areas cannot afford computers or do not have access to electricity, let alone an Internet service provider, but they can use sites maintained by other organizations, communicating by mail. Though they cannot afford printing, they are still able to reach out to other interested people the world over. The Zapatista movement in Mexico has become particularly famous for its use of the Internet, but there are many more.

At a local level, the older methods are still the most effective: political street and village theatre provides some of the very best satire; musicians have always had a large part to play in communicating criticism and keeping up spirits. Particular dress codes or hair styles, gestures such as freedom handshakes, use of a regional language or even cuisine can all be ways of communicating. All resistance movements quite openly recognize that the cultural front is the most important one on which they operate; without it, nothing else can succeed.

SUBVERSION

Emerging from this we can see that many cultural geographers have become interested in cultural counter-hegemony. Several of these would describe themselves as anarchist, rather than Marxist. Their work can include studying the ways in which groups subvert established forces. Whatever one's politics, it is interesting to observe the ways in which excluded people reject their exclusion, ranging from legally sanctioned demonstrations, publicity and exhibitions to invasion, vandalism and theft. Tim Cresswell's book *In Place/Out of Place: Geography, Ideology and Transgression* (1996) has been very influential in making subversion a recognized area of study, though the inspiration is poststructuralist rather than Marxist.

Often the politics of resistance takes the form of spectacle, shock or irreverent play as an effective means of subverting power. Mitchell (2000) considers spectacular and carnivalesque activities as protest; he sees the events of Paris in May 1968 as the point when students in revolt appropriated the forms of spectacular performance, becoming semi-theatrical and artistically self-conscious in their control of the streets. Their slogans are unforgettable ('Be realistic, demand the impossible', 'It is forbidden to forbid', 'Defend the collective imagination') but their surrealism was incomprehensible to many old style left-wingers. For all the playfulness, the intention was serious and the Paris events had reverberations for youth and student culture as the protests spread throughout the Western world. Age, like gender and race, began to be recognized as another major category of economic and political differentiation which modified the ways in which one interpreted the connection between class and culture.

Peter Jackson (1988a) has shown how carnivals themselves are a means whereby racial (and other) minorities can claim recognition and respect. Tim Cresswell cites the gay activist 'Act Up!' marches and events. Rob

Shields (1991) refers to staged fights in Brighton between gangs of young people from London to consider how marginal places can be used as what he calls 'carnivals of violence'. Recently several younger cultural geographers (Skelton and Valentine, 1998) have carried out research on the resistance aspects of contemporary youth cultures; this can be a fascinating field for undergraduate dissertation work, since the youth of the researcher is an unambiguous advantage to both the effective collection of information and its sensitive interpretation.

TASK

> • Unless you are of a nervous disposition, walk through your town centre on a Saturday night.
> • Who is on the streets? (And who is not?)
> • How are they behaving?
> • Would it stretch your credibility to call what you see 'resistance'?
> • Resistance to what?

There are obvious connections between a critique of society based upon Marxist ideas relating to structure and one which relies on deconstruction. Both question the basis of prevailing structures of power and knowledge. It has always seemed a pity that those who criticize the existing order, and the cultural devices used to perpetuate it spend such a disproportionate amount of time trying to demolish the arguments of other critics with a different perspective. There is often a deep distrust between Marxist and post-structuralist approaches to contemporary culture (Harvey 1989b; Peet 1998), but regrettably a 'third' view which attempts a reconciliation can seem equivocal (Soja 1996).

Perhaps cultural geographers who reject Marxism might find it useful to grapple with the new value system underwriting the capitalist utopian philosophy of van Parijs (1995). Van Parijs challenges existing assumptions about capitalism and socialism to construct a world of maximum freedom which is combined with maximum equality of opportunity, accompanied by an unconditional entitlement to a basic income (regardless of unequal abilities or willingness to work). His philosophy requires his readers to question foundational moral issues, such as the work ethic, whilst contemplating what constitutes a just society. I find that there is no better way of revealing the problematic nature of one's own ideology (whatever it is) than to confront what seem like outlandish views. Certainly van Parijs' capitalism is not the familiar sort.

Engagement with Marxist or post-Marxist thought assumes a degree of commitment to those one believes to be exploited or misrepresented. It is not a detached intellectualist position but relates itself to the world. Most obviously this can mean involvement with groups working for various issues (Routledge 1996), but 'doing' Marxist cultural geography does not invariably have to imply activism, despite Marx's observation that it is change rather than mere understanding that should be the focus of study. Marxist thought can be applied to any aspect of cultural geography. There is a tradition of Marxist literary criticism which can inform the interrogation of texts with geographical relevance. Ahmad (1997b) advocates reading Arundhati Roy politically, and criticizes the way in which Roy's novel *The God of Small Things* was written with a Western anti-communist readership in mind. He attacks not just the content of her story but also its form as pandering to Western tastes, even though he applauds her skill. Many other cultural products, not just books, can be read against the grain.

GLOSSARY

Commodity   Thing, service, experience which is produced for profit.

Fetishism   Elevating a thing beyond its use value (a fetish was a West African religious object).

Hegemony   Domination (cultural hegemony, domination of cultural processes).

FURTHER READING

Harvey, D. (1989) *The Condition of Postmodernity: an Enquiry into the Origins of Cultural Change*. Oxford: Blackwell.
Harvey, D. (2000) 'Reinventing geography', *New Left Review*, Second Series, 4: 75–97.
Mitchell, D. (2000) *Cultural Geography: a Critical Introduction*. Oxford: Blackwell.
Peet, R. (1998) *Modern Geographical Thought*. Oxford: Blackwell.

 # 5 Poststructuralist cultural geography

Pamela Shurmer-Smith

It is interesting to consider why poststructuralist thought so often attracts vitriolic criticism (and easy sneering) from both the ideological left and right. My own sympathies are to the left, but my reading of Marx makes me feel that, in trying to conceptualize the rapid economic, political and social transformations in the world today and the manner of their communication, he too would have understood and tolerated poststructuralism.

> Constant revolutionising of production, uninterrupted disturbance of all social conditions, everlasting uncertainty and agitation distinguish the bourgeois epoch from all earlier ones. All fixed, fast frozen relations, with their train of ancient and venerable prejudices and opinions, are swept away, all new-formed ones become antiquated before they can ossify. All that is solid melts into air, all that is holy is profaned, and man is at last compelled to face with sober senses his real conditions of life and his relations with his kind.
>
> (Marx and Engels 1960 p. 18)

It is, however, difficult to theorize things which are in a state of flux. People who are reluctant to jettison ideas that have served them well in the past find it difficult to come to terms with the insecurities of theorizing rapid change, whereas structuralist thought requires an assumption of some degree of fixity (including an assumption that elements can stay more or less the same between generations through processes of social reproduction). Contemporary capital is, however, volatile and apparently fickle; one now expects change rather than stasis. Everywhere in the world, one generation is unsure how best to reproduce the next in terms of the global division of labour or in terms of new mediatized culture. When flexibility of skills and perspectives becomes valued more positively than traditions and certainties, the most useful frame of mind is, surely, the ability to conceptualize and live with change. Poststructuralism encourages one to think the processes of becoming, rather than states of being, and, though it is regularly accused of being mere fashion-chasing, most of us know in our hearts that the old certainties are never coming back. Many of us would not want them to; they are defined by categories that were not of our choice. Anyone who is not white, male, affluent, heterosexual *and* conventionally well educated in Western society has little reason to conserve old structures of life or thought. To take an example close to student experience, it is instructive to consider the reasons for the late arrival of issues of race and ethnicity, popular culture, gender and sexuality on academic curricula. Many conservatives still do not think these worthy of serious study.

We live in a concrete material environment and we share basic biological, social, intellectual and perhaps even spiritual capacities; we also share the capacity to reason. Losing site of this basic reality comes from too great an emphasis on difference and diversity – an extreme position that can be found at the margins of postmodernism, multiculturalism, and radical feminism. These go beyond criticizing reason for not living up to its own standards. They urge us to remove reason from any position of privilege.

(Sack 1997 p. 4)

Poststructuralism recognizes its descent from structuralism and it can be argued that, rather than being opposed to structuralism, it accepts and goes beyond it, retaining all of structuralism's romantic and positivist enemies and their descendants too. Poststructuralism attempts to break down conceptual structures by questioning the basis of their construction, but it also seeks to discover how one might theorize formations which will never have time to harden into structures.

It is hardly surprising that such a field of endeavour would attract charlatans and poseurs, as well as serious thinkers struggling to generate new ways of conceptualizing. The category *postmodern* has become virtually meaningless through overuse; it is generally associated with difference for its own sake, playful mixing of genres and deliberate shock tactics. It has much going for it as a style which brings home to people in establishment positions in the West the realization that 'the times they are a'changing', but many of us believe that it is necessary to do more than trumpet the death of the metanarrative (Lyotard 1984) and to 'celebrate' difference as if at a fancy-dress party. (The phrase 'celebration of difference' was not actually intended to imply festivities: it just means 'making difference known', as in *cause célèbre*, but few seem to know this and the meaning has wandered.)

> [The] extreme version of postmodernism propagated in the name of intellectual anti-terrorism employs the terroristic device of terminological vagueness, attempting to overpower the reader with indecipherable phrases and a 'quotation' market of its own, this time from Derrida, Deleuze, Guattari and others, that only the privileged can afford to read (because of the preparation needed) . . . Thus it has a politics but one of nihilistic terror.
>
> (Peet 1998 p. 242)

It was almost inevitable that poststructuralism would emerge in the fragments of academic life which specialize in understanding culture – literary criticism, architecture, cultural studies, cultural anthropology; it was also inevitable that it would be espoused by many of the 'new' cultural geographers. The processes of textual deconstruction migrated into many different areas of study, collapsing the artificial boundaries which had created and maintained disciplinary specialization. The metaphor of the *text*, which could be interpreted, took the place of the *building*, which could be taken apart.

As suggested above, the people most attracted to deconstruction were (and are) those who were aware that they had been constructed as marginal, eccentric, excluded from the main project. Notions of alterity, the transgression of boundaries, the questioning of categories, thinking 'against the grain', experimental modes of communication and a degree of irreverence for hallowed institutions became characteristic of poststructuralist thought. This was not, as denigrators have suggested, because of a lack of seriousness of intent: it was a means of discovering the frailty of many things taken for granted at the centre.

TASK

- Have you ever felt that an interpretation you have read in a geography text excluded the way you perceive the world?
- Can you explain this?
- Think also about your fellow students: which people find it easiest and hardest to relate to 'normal' academic conventions?
- Why do you think this is so?
- Now decide how you think that ideas about intelligence and stupidity emerge.

The language of deconstruction is inherently geographical: it utilizes imagery of spaces, flows, intensities, contiguities and (as geographers have always done) contemplates how things have come to be situated. These questions are not exclusive to this approach but poststructuralism has made it easier for cultural geographers to progress beyond simple description, essentialist senses of uniqueness or determinist notions of construction of places. It has allowed place and ideas of emplacement to be freed from assumptions of fixity, incorporating relativity and flux into geographical discourse.

Most poststructuralist thought originates in France, growing out of the upheaval of French universities following the events of May 1968. Some of the unfamiliarity of poststructuralist writing emerges from the fact that it is encountered by English readers in translation. French has different conventions of expression, it is more wordy than English and it values word play; translators who have attempted to capture this in English have generally managed to make it sound more ponderous and pretentious than was ever intended. The point of reading poststructuralist philosophers is not to learn their every detail but to glimpse alternative ways of thinking.

MICHEL FOUCAULT

Foucault has probably had more influence on contemporary geography than any other poststructuralist thinker. His views about the movement from premodern to modern societies being marked by a shift from a reliance on punishment to (self-) discipline as a form of social control appealed to historical geographers interested in the qualitative changes in European institutions. Foucault's reliance on the archive (hence his archaeology) as the source of traces and remnants of a general history have struck a chord with radical historical geographers, and will be explored in more detail by Kevin Hannam later in this book. Felix Driver (1985) is responsible for alerting geographers to the ways in which Foucault's sense of the relationship between power and knowledge might be incorporated into a geography which was sensitive to the disciplinary use of space. Chris Philo (1992) further refined the ways in which 'Foucault provides us with . . . a blueprint for a truly "postmodern" geography in which details and difference, fragmentation and chaos, substance and heterogeneity, humility and respectfulness, feature at every turn, and an account of social life which necessarily brings with it a sustained concern for the geography of things rather than a recall for the formal geometries of spatial science' (1992 p. 137).

It is from Foucault, in particular, that geographers have derived their recent interest in *discursive practices* (the construction of bounded knowledge systems which encourage thought along approved lines) and *surveillance* as the means of control of people and practices. The gaze is an expression of power, as indicated in the school teacher's threatening 'I've got my eye on you', or the aggressive 'What are you looking at, then?' Foucault's thinking made it difficult to refuse to separate cultural and political constructions from each other and from the manipulation of physical entities in space.

TASK

> Foucault's ideas about surveillance come alive when you use them to think how people manipulate others by controlling what they can see and how they are seen.
>
> - How do people you encounter use space to enhance their own power? Think about teachers, officials, hospitals, shops, concerts and so on.
> - Look around your everyday environment for CCTV cameras and decide what they are intended to achieve in different places.
> - Think what other means of surveillance you are subject to. What is being controlled and by whom?
> - Think about the trade-off between privacy and safety. What is your attitude? What do other people think on the matter and how much does their experience influence this?

TASK

> Foucault's ideas about discourse gain some immediacy when you use them to think about the different ways people write and speak and how these do not just communicate ideas, but also restrict what can be communicated.
>
> - Try rewriting an academic essay using no technical terms but employing the language you use when speaking to your friends.
> - Try writing a love letter in the style and register recommended for writing up a scientific experiment.
> - Think why giving a formal presentation is so often a harrowing experience.

GILLES DELEUZE AND FELIX GUATTARI

Although both also wrote independently, geographers know Deleuze and Guattari best for their joint work, particularly *A Thousand Plataeus* (1988), the second volume of *Capitalism and Schizophrenia*. It takes a leap of faith to read this book, because it abandons virtually all of the conventions of academic writing, but it is worth persisting with because one eventually becomes familiar with their style and their radical approach. The authors say that the chapters can be read in any order but that one should read the conclusion last; I'd suggest that you read a few chapters and, once you are throughly confused, read the conclusion, then have another look at the chapters.

One finds in Deleuze and Guattari many of the devices familiar in contemporary art and media, particularly film, advertising and music. They emphasize wandering (nomadism), fragmentation, flash of insight, volatility, radical rupture, the repression of the mass: reading Deleuze and Guattari is like experiencing some wild disorienting music mix, slightly frightening but liberating.

Let us summarize the principal characteristics of the rhizome: unlike trees or their roots, the rhizome connects any point to any other point; and its traits are not necessarily linked to traits of the same nature; it brings into play very different regimes of signs, and even nonsign states. It is not the one that becomes two or even directly three, four, five etc. It is not a multiple derived from one, or to which one is added ($n + 1$). It is composed not of units but directions.

(Deleuze and Guattari 1988 p. 21)

Deleuze and Guattari are making the point that structured knowledge utilizes the metaphor of the tree, with its roots and branches, one bifurcating from a higher-order other. Much explanation in geography has conventionally taken this systems approach. They are suggesting that if one takes another plant metaphor and thinks in terms of the rhizome one has a non-bifurcating, spreading view of knowledge where growth can potentially come from all points.

Smooth space and striated space – nomad space and sedentary space – the space in which the war machine developed and the space instituted by the state apparatus – are not of the same nature. No sooner do we note a simple opposition between two kinds of space than we must indicate a much more complex difference by virtue of which the successive terms of the oppositions fail to coincide entirely . . . the two spaces in fact only exist in mixture: smooth space is constantly being translated, transferred into striated space; striated space is constantly being reversed, returned to smooth space. In the first one organizes even the desert; in the second the desert gains and grows.

(Deleuze and Guattari 1988 p. 16)

Here they are referring to the way in which structures are constantly being worn down, but conversely structures impose themselves where there are none. This is an important part of their argument, since it shows that poststructuralist thinking does not deny structures (or their inevitability), it merely questions the stability of their form.

> *Break line, crack line, rupture line.* The line of rigid segmentarity with molar breaks; the line of supple segmentation with molecular cracks; the line of flight or rupture, abstract, deadly and alive, nonsegmentary.
>
> (Deleuze and Guattari 1988 p. 200)

'The line of flight' is a radical break from old forms. It contrasts with other metaphors of change: 'rigid sementarity' where forms divide down into component parts (e.g. nations dividing into sub-state nationalisms), and 'supple segmentation' where a system changes through weakening and bending (e.g. a political system staying the same in form but becoming corrupt in practice).

TASK

'Doing' is all very admirable, but don't try writing assessed essays in imitation of Deleuze and Guattari! (Things have changed in academic life, but not that much.) The best use you can make of them is to employ their metaphors to problematize some of the conventional ways of thinking about space.

- Think about the instances where the tree is used as a metaphor in systems approaches to knowledge. See what happens when you substitute the rhizome.
- Does the 'line of flight' duck the issue of how change occurs?

JEAN BAUDRILLARD

Though Baudrillard began writing as a Marxist interested in commodity fetishism, he moved away from Marxism completely to contemplate value removed from substance. The later Baudrillard is often cited as having abandoned morality, or even reality (this does not make him any less interesting). One needs to be careful to check the dates of original French publication of Baudrillard, as many of the earlier works were translated into English only after his more extreme later works became popular; if one went by the English dates, one would experience a very strange progression of thought. For example *The Consumer Society* (Baudrillard 1997) which laid down important ideas about the way in which consumption was more important than production in a post-industrial society, was originally published in 1970 but reached monolingual English readers very late, in 1997, well after English readers had become familiar with radical expressions of hyperreality. His early works throw much of his later 'extremism' into perspective and make it more valuable.

Though it is difficult to admire some of Baudrillard's posturing (for example, I find the jottings *Cool Memories II* (1996) hard to take seriously, or even to enjoy frivolously), he has been crucial to the way in which cultural geographers, amongst others, think about the mediatized experiences which dominate life in the West. Most importantly, Baudrillard introduces the idea of the *simulacrum*, the simulation which has become divorced from any 'real' thing, but which has become a phenomenon in its own right. So, for example, one visits Disneyland, a fake place which has become real, to meet real actors impersonating cartoon characters; whilst there, one can go on risk-free rides which will make risk-filled reality seem tame.

> Once, out of some obscure need to classify, I proposed a tripartite account of value; a natural stage (use-value), a commodity stage (exchange-value), and a structural stage (sign-value) ... let me introduce a new particle into the microphysics of simulacra. For after the natural, commodity and structural stages comes the fractal stage ... At the fourth, the fractal, or viral, or radiant stage of value, there is no point of reference at all, and value radiates in all directions, occupying all interstices, without reference to anything whatsoever, by virtue of pure contiguity. At the fractal stage there is no longer any equivalence ... properly speaking there is no law of value, merely a sort of *epidemic of value, a sort of metastasis of value, a haphazard proliferation and dispersal of value.* Indeed we should really no longer talk of value at all, for this kind of propagation or chain reaction makes all valuation impossible.
>
> (Baudrillard 1993b p. 5)

Hyperreality is more real (sensation packed) than reality. In a highly mediatized world, the real is liable to become a disappointment in comparison with the perfect fake; the hype and the hyper become the source of value. Baudrillard alerts us to the dangers of seeing the world only in mediatized forms; he goes beyond the obvious warning that the medium does not tell the whole truth to confront us with the way in which the hyperreal escapes into lived experience to corrode it. Cultural geographers have particularly drawn upon Baudrillard for an understanding of the consumption of postmodern images, particularly environments such as theme parks, holiday villages, world fairs, shopping malls, hotels and restaurants where the fakery is not just blatant, but is presented as a major part of the enjoyment. Baudrillard goes well beyond mourning for the loss of authenticity (as, for example, in the Frankfurt School); he seeks, rather, to discover the power of the hyperreal, which we can see as being super-inauthentic.

None of this is trivial. In those parts of the world where most people are fortunate enough to be able to satisfy all of their material needs and most of their material wants, the greatest profits are to be made in stimulating desire through manipulation of fantasy. As I write, a hyperreal rage is sweeping through the school playgrounds of the affluent world. I refuse to give extra free advertising by naming the phenomenon whereby children, having seen a feature-length cartoon film and regular TV cartoons, which are spin-offs from a computer game, have been persuaded to believe that cards with kitsch pictures of 'monsters' are both valuable and scarce, but I am sure that you know what I am referring to. They compete to collect full sets by buying and swopping but also by stealing and cheating. Even a killing has been reported: did it really happen, or is it just another cynical part of the hype? Only a concept of the hyperreal and the notions of fractal and viral value begin to make any sense of this and its adult equivalents.

TASK

> Baudrillard's claim that 'there is now no law of value, merely a sort of *epidemic of value*' (1993b p. 5) is one which can offer a useful starting point for cultural geographers who want to go beyond mere description of postmodern urban environments.
>
> • Think of some examples of epidemics of value you have experienced.
> • If there are epidemics, can there be an epidemiology of value? How might this be realized?
> • If there are viruses, can there be antidotes?

HÉLÈNE CIXOUS

Cixous is pre-eminent among the women of the poststructuralist movement because it is to her that we can attribute the idea of *l'écriture feminine*, the proposition that it is possible to discover a mode of writing in the feminine, avoiding what Lacan revealed as *phallogocentrism*, writing from the assumption of masculine power. Taking Cixous' perspective, one realizes that it is impossible for an academic woman to write from her own point of view; the language simply does not permit it (how could a woman *master* her subject; how can she keep a straight face if she thinks about the metaphor underwriting the *seminar*?)

Cixous is concerned about the walls that are constructed to exclude the feminine from the centre; she demands that people recognize that the only reason for this exclusion is in the body, and (along with Foucault) she brought the body into the supposedly purely cerebral arena of academia. Her insistence on the embodied nature of thought is unsettling and is taken further by Luce Irigaray. She shies away from conventional thought based upon binary oppositions because she believes that, in every structural opposition, the one must necessarily take precedence over the other and, in doing this, continuities are obliterated.

Cixous works to reveal the oppression that is contained in oppositional thinking, whether this be along the lines of gender, ethnicity or age; and this causes her, like Deleuze and Guattari, to emphasize flow, continuity and the process of fragmentation. She takes the risk of writing her 'academic' works in non-academic style but simultaneously writes theatrical plays which seem obscure and intellectualized. For cultural geographers the interest in her work emerges from the deconstruction of binary opposites, but also from her use of space as if it were a tangible element in human interactions (Shurmer-Smith 2000a). Cixous eschews place: she writes of Algeria, but it never takes on tangible qualities. For Cixous space is far more important, particularly the way that people can dominate space or are enfeebled by the spatial strategies of others.

Cixous, like Derrida and Camus, was born and grew up in Algeria, France's most problematic imperial possession (white Algerians, known as *pieds noires*, were destined always to be outsiders in metropolitan France, whatever their personal politics). She always found it difficult to accept the constructions which threw her to the margins – Jewish, Algerian, woman, poor – and has dedicated much of her writing to constituting a seamlessness which denies these constructions. She is passionately anti-racist and anti-nationalist, but seeks to understand the ways in which essentialist identities are forged. This sometimes makes her appear to be writing from an essentialist perspective, which she most definitely is not, even though, particularly in her plays, she takes an epic stance. As one might expect, her view of nature is not separate from humanity and in her novels the human often merges directly into the natural, in a way which would be disconcerting for a geomorphologist.

Like many poststructuralists, Cixous adopts the stance of the traveller, not fixed in time, space or social categories. She urges her readers to accept the fluidity of definitions and the realization that stasis is illusionary.

DERRIDA

Derrida has had less direct influence on geographers than many of the other poststructuralist theorists and his work is often difficult to read in its cold intellectualist attention to language. He is, however, important to cultural geographers for alerting us to the importance of the text, not just for an understanding of the intentions of the author, but also for unlocking the interpretive faculties of the reader. Derrida draws attention to what is *not* written, the words and ideas that are discarded or never considered when a text is being constructed, the things which are deleted from each writer's world view. After Derrida it became normal for cultural geographers to ask why certain people, things, ideas, places were 'missing' from a representation. Obviously this was vitally important to the growth of studies from minority perspectives such as those of postcolonial, feminist, sexuality theorists ('minority' is used not just in the numerical sense, but to mean being minor in terms of power).

THE STATUS OF POSTSTRUCTURALIST THOUGHT IN GEOGRAPHY

There was great enthusiasm for poststructuralism in the cultural geography of the early 1990s, but it is certainly declining in popularity. The research group in cultural geography at the University of Portsmouth started out with the noble intention of showing how poststructuralist theory could inform the practice of ethnographic research, not just the writing but also the fieldwork. One by one the members of the group moved to more conventional ways of doing things, perhaps because it is difficult to *be* a poststructuralist.

The work of poststructuralist theorists, however, is a continuing influence and has been accepted into mainstream thought, as in David Sibley's sensitive use of Julia Kristeva's ideas about abjection to understand the geographical outcomes of processes of stigmatization and exclusion. The main value of deconstruction lies in its critical powers and its message that things do not have to be the way they are, or be explained in ways which maintain existing structures. The big question in poststructuralism is one which should underlie all intelligent geography, 'Why is it this way?'

GLOSSARY

Alterity   Difference (as a distinguishing feature).

Discourse   Language practices which structure thought and behaviour.

Eccentric   Coming from the margins of what is regarded as normal.

Hyperreality   Beyond reality (better than real).

Simulacrum   A fake which does not have a real thing behind it (a simulation of a simulation).

Surveillance   Control by means of monitoring.

FURTHER READING

Harvey, D. (1989) *The Condition of Postmodernity: an Enquiry into the Origins of Cultural Change*. Oxford: Blackwell.

Sarup, M. (1993) *An Introductory Guide to Post-Structuralism and Postmodernism*. London: Harvester Wheatsheaf.

Soja, E. (1989) *Postmodern Geographies: the Reassertion of Space in Critical Social Theory*. London: Verso.

6 Feminist cultural geography

Carol Ekinsmyth

Since the early 1980s, feminist geographers have engaged in a project to recast the 'centre' of the discipline of geography and to reconstruct knowledge, through advocating different ways of defining subject matter, different ways of knowing and different ways of 'doing'. The feminist critique of geography (and of science more generally) can be crudely understood as a statement of dissatisfaction with modes of enquiry that assume a male, white, heterosexual, middle-class, middle-aged vantage point and, in so doing, exclude or alienate those who do not belong to this group. The critique began in geography with a call to recognize gender as a fundamental social category and to include women in the subject (both as those studied and as producers of knowledge). Monk and Hanson, in one of the earliest calls, aptly titled their 1982 paper, 'On not excluding half of the human in human geography'. The debate and arguments have intensified and become more complicated over the intervening 20 years, although feminists have never lost sight of their political project to improve the lives of women. This chapter will discuss some of the debates in feminism and feminist geography that have made an important contribution to cultural geography generally, and feminist cultural geography in particular.

> One expression of feminism is the conduct of academic research that recognizes and explores the reasons for and implications of the fact that women's lives are qualitatively different from men's lives. Yet the degree to which geography remains untouched by feminism is remarkable, and the dearth of attention to women's issues, explicit or implicit, plagues all branches of human geography. Our purpose here is to identify some sexist biases in geographic research and to consider the implications of these for the discipline as a whole ... we would argue that, through omission of any consideration of women, most geographic research has in effect been passively, often inadvertently, sexist.
>
> (Monk and Hanson 1982 p. 11)

Indeed, just as there are multiple versions of feminism – feminisms rather than feminism – so too are there of feminist geography. And since this book is an exploration of these various feminist geographies, then in many senses we want to leave the task of unravellling these various feminist geographies to you. One point, however, on which we are all agreed, is that notwithstanding the existence of feminist geographies, the practice of feminist geography is one which contests and challenges the frequent taken-for-grantedness of the content and concepts which sit at the heart of the discipline of geography, and their assumptions.

(Women and Geography Study Group 1997 p. 5)

TASK

- Make a note of the similarities and differences between the two extracts given.
- What do these tell you about the origins and development of feminist geography?

There is much overlap of opinion, objectives and method between feminism and poststructuralism. Feminism is more than a concern with and for women, but women are one of the categories which Pamela Shurmer-Smith describes in Chapter 5 as having been 'constructed as marginal, eccentric, excluded from the main project'. For this reason, many feminist researchers have been attracted or compelled to the poststructuralist and feminist project of challenging and destabilizing the 'centre'. Aside from Hélène Cixous who has been discussed in Chapter 5, Gayatri Chakravorty Spivak's work is often cited as the best and most geographical example of poststructuralist work in a feminist vein (McDowell and Sharp 1999; Spivak 1988b). By no means all feminist geographers, however, work in a poststructuralist vein. Early feminist geographical work addressed the project of adding women into geography (now often referred to as the 'add women and stir' method), where geography as conventionally practised had its scope broadened to include women's lives and women's activities (e.g. the home as a sphere of geographical interest, unpaid labour, geographies of child care). It was not long, however, before feminist geographers began to question whether there might be more appropriate epistemologies (ways of knowing) and ontologies (ways of being or existing) for feminist endeavour. Science, social science and scientific enquiry have by and large been created by men. Men have been the legislators of what counts as legitimate knowledge, what counts as acceptable 'proof' of existence; feminists began to wonder what women's knowledges would be like, how they would differ, how a science created by women would be. Some have bravely (and it is brave, since conventional science and the academy have the power to reject and ridicule this work and break careers) stepped out along the path to creating one. Others have continued to work within more orthodox frameworks, expanding the scope of these frameworks and reinforcing the feminist message that women count.

It is not surprising, given feminism's concern to expose and understand those systems and processes that lead to inequality between women and men, that feminists have been drawn in large numbers to the 'new' cultural geography. Many feminist geographers have also become cultural geographers, and their work has contributed to the project through drawing attention to the ways in which gender relations are inscribed in modes of representation and systems of cultural reproduction. Seeking an understanding of the power relations that this entails and legitimates is integral to the project.

GENDER, DIFFERENCE AND DIVERSITY

Feminists have highlighted the need for gender to constitute a central feature in cultural geography. In understanding modes of representation and systems of meaning, it is vital to recognize that gender is constitutive of and constituted by these modes and systems. In concrete terms, for example, gender roles and relations are born out of everyday practices, and in turn, the nature of these everyday practices is in part the result of taken-for-granted understandings of gender. This mutually constitutive relationship has been the focus of much feminist cultural geographical research, in wide-ranging empirical contexts (see for example Laurie et al. 2000). Feminists would argue that a cultural geography that does not acknowledge the centrality of gender is an impoverished cultural geography.

Over the past 10 years or so, however, feminists have had an uneasy relationship with 'gender' and the category 'woman'. The earliest work in feminist geography tended to treat 'gender' as an unproblematic concept, producing work that represented women as a homogeneous group. Critiques from black feminists who argued that this work could not speak for the experiences of black women, and lesbian feminists who objected to the implicit heterosexuality in most feminist writings, as well as the growing influence of poststructural and postcolonial researchers who questioned monolithic categorizations such as gender and binary opposites such as 'woman/man', began to dislodge feminism's faith in the stability of gender as a construct (hooks 1992; Spivak 1988b; Lorde 1984; Valentine 1993). There is now theoretical acceptance amongst feminist geographers that the identity of any individual is made up of multiple and shifting positions (gender, sexuality, race, class, religion, age, nationality, wealth etc.) and that there is thus no monolithic category 'woman' that can be studied or represented in this way (hooks 1984; 1991; Rose 1993; Valentine 1997b;

McDowell 1997b). Feminist cultural geographers have instead engaged in exciting work to understand the social construction and the shifting and performative nature of identity. Gender is conceived, not as a static identity position that an individual acquires through early stages of development and socialization, but instead as an ever shifting set of possibilities and performances in any given context. 'Gender, it is argued, is constructed and maintained through *discourse* and everyday actions' (McDowell and Sharp 1999 p. 108, emphasis in original). Location matters in this respect: 'gender' performed in paid work might be very different from 'gender' performed in the home or 'gender' performed in a night-club. The culturally endorsed rules for these 'performances' restrict and control their nature. The focus of attention for those interested in understanding gendered experiences is broadened through these insights, to include discourses, regimes of representation, modes of control and everyday actions (see for example Jackson 1991 on masculinity). For geographers, specific places are also the focus of attention as forums for the creation and maintenance of gender identities.

On Merchant Banks

I have demonstrated here, and in the previous chapter, how women are made to feel out of place on the trading floors and in the dealing rooms [*of merchant banks*] by the development of a particular type of heterosexual machismo culture in which crude bodily humour, pin-ups, practical jokes and various forms of verbal and non-verbal behaviour verging on sexual harassment are the norm. The bodily imagery that is commonplace in the everyday language and social practices of trading and dealing rooms relies on a particularly exaggerated version of masculinity and masculine performance. The most extreme example is the dubbing of successful traders as 'big swinging dicks'. Power, sexuality, desire and masculinity combine to construct femininity as deficient.

(McDowell 1997a pp. 178–9, italics added)

Just as gender is understood by feminists as a social construction, following the persuasive arguments made by Judith Butler (1990), and influences from poststructural and postmodern theorists, many feminists have also come to understand 'sex' as a social construction too. Butler argued that the view that there are two natural sexes, defined by biology and manifested in the physical qualities of bodies, is a fiction. Just as gender is a 'regulatory fiction', something which is learned and performed rather than something which is 'natural', so too is 'sex' .

> Sex is what makes our bodies male or female, but what is sex? What is it made up of? Is it made up of genes, genitals or hormones? Is a woman without a womb still female? Is a woman injecting herself with testosterone still a woman? Is a man carrying the chromosomes XXY still male? Perhaps sex cannot be reduced to biology. Perhaps there is no such thing as 'natural' sex? The notion that we all fit into either a male or female body is just that, notional. It cannot be sustained over either time or space. Simply put, there is nothing 'natural' about 'male' and 'female' bodies.
>
> (Women and Geography Study Group 1997 p. 195)

As the Women and Geography Study Group (1997) reminds us, children born possessing bodies with qualities associated with both sexes are often surgically 'corrected' to turn them into one sex or the other. Similarly, the Group reminds us that womanhood is a fiction, and refers to the huge amount of effort women have to make to maintain their 'feminine' bodies: hair removal, skincare, dieting, the wearing of feminine clothes and shoes that make them 'walk like a woman', even cosmetic surgery. Reproductive technologies also call into question conventional understandings surrounding sex.

These realizations are very exciting for feminist cultural geographers who have begun to focus on the regulators and reproducers of these culturally specific fictions. Just as gendered performances have been the focus of attention, these insights about sex bring the body more fully onto the agenda in geography. Embodiment and embodied performances in sexed spaces and places are the focus of new work, as are the mechanisms of the control of these performances (see for example McDowell 1997a; Lewis and Pile 1996; Pile 1996; Parr and Philo 1995). The body has been redefined as a geographical scale worthy of attention by geographers (Duncan 1996). Related to this is an interest in sexuality which will be discussed in a later section of this chapter.

The problem for feminists in accepting the multiple and shifting nature of identities is that the political project of improving women's lives is difficult if there is little acceptance of the category 'woman'. Here, feminists have entered a debate over 'essentialism'. In simple terms, essentialism is the attribution of characteristics to individuals on the basis of their inherent characteristics. Thus the common belief that women are more emotional or more dextrous than men is an essentialist belief. Similarly, the belief that inequalities of brain power or ability between women and men are due to biological difference is essentialist. Equally, an understanding of woman as a socially constructed category which is never the less tangible is also essentialist, as it attributes characteristics or subject positions to women on the basis of their female bodies. Feminists have been caught in a dilemma over essentialism. It is clear (and decades of feminist work has proven) that women as a category are positioned differently and unfavourably *vis-à-vis* men. The feminist project aims to eradicate these inequalities, but how can this be achieved if the analytic and theoretical shortfalls and the inherent dangers of essentialism are acknowledged, to the extent that we cannot speak of 'women' as a category? If we concentrate on difference and diversity amongst women, how can we effect political change? Audrey Kobayashi has referred to the debate over essentialism as 'perhaps the most exciting but painful crisis in feminism' (1997 p. 7).

Many feminists believe that a strategic use of essentialism ('strategic essentialism': Fuss 1989; de Lauretis 1986; Kristeva 1984) is necessary for an effective feminist politics. Whilst acknowledging that women are different from each other, it is never the less the case that many of the experiences that women have in male-dominated societies lead to commonalities amongst them. If this is questionable, feminists have no cause to fight for. Thus, as a political device, strategic essentialism – the investigation of women's lives, and thus the treatment of 'women' as a group – is entirely warranted (though contrast Kobayashi 1994).

THE GENDERING OF SPACE AND PLACE

Feminist geographers have shown that both space and place are gendered. Indeed they have gone further than this to argue persuasively that gender and space are constitutive of each other. As Laurie et al. (2000) argue and empirically demonstrate, 'gender and space constitute and are constituted by each other in two ways: gendered identities are constituted in and through particular sites; these sites are constituted through our ideas about gender' (p. 162).

TASK

> • Think carefully about what the previous sentence means.
> • Now think of some 'sites' where gendered identities are constituted.
> • How are these sites, in turn, constituted by societal ideas about gender?

This is a key insight in our quest to understand how gender identities are created and also what spaces and places mean to people. Feminists have long argued that in the Western world, day-to-day life (the economy, society and the form of the built environment) is structured around cultural understandings of 'public' and 'private' as well as gendered understandings of space and place. Women's 'natural' sphere or place is often thought to be the private sphere of home, family and residential areas (typically the suburbs) whereas men are thought to belong in a public sphere of paid employment, public space and the central areas of cities. Of course, the separation of men and women into these spheres is not complete, and daily experience tells us that women are found in great numbers in the so-called public sphere of paid work and city centres. But the pervasive myth is familiar to us in all kinds of representational forms from advertisements to film, magazines, novels and television programmes.

TASK

> - Find some specific examples of representational forms (e.g. film, magazine articles, advertisements) that reinforce pervasive myths about a 'woman's place'.
> - List the ways in which they do this.
>
> (In some cases, messages will be blatant; in others, you may have to look for subtleties which are none the less powerful.)

Feminist cultural geographers have studied the messages contained within these representational forms, and they have argued that these under-standings structure lives, opportunities, city planning decisions and the design of urban space.

> The city – as experience, environment, concept – is constructed by means of multiple contrasts: natural, unnatural; monolithic, fragmented; secret, public; pitiless, enveloping; rich, poor; sublime, beautiful. Behind all these lies the ultimate and major contrast: male, female; culture, nature; city, country. In saying this, I am not arguing (as do some feminists) that male–female difference creates the deepest and most fundamental of all political divisions. Nor am I arguing that the male/female stereotypes to which I refer accurately reflect the nature of actual, individual men and women. In the industrial period, nonetheless, that particular division became inscribed on urban life and determined the development and planning of cities to a surprising degree and in an extraordinary unremarked way.
>
> (Wilson 1991 p. 8)

Elizabeth Wilson (1991) argues that the shape of contemporary cities results from underlying assumptions of men's and women's roles and places. Dolores Hayden (1981), Jos Boys (1998) and Clara Greed (1994) have also written about the built form of the urban environment and its importance in the production and reproduction of masculinist societies. In a nutshell, the argument is that the zoning of land use in cities, with the separation of paid employment from residential areas, makes concrete the expectation of a gender division of labour (particularly that 'a woman's place is in the home').

TASK

- Think how and why this could be so.
- In material terms, what are these feminists arguing?
- Think about who you would expect to be in the following public spaces: residential suburban streets at 12.00 noon; walking across Waterloo Bridge in the city of London at 8.00 in the morning; at an out-of-town shopping centre at 3.30 p.m. on a weekday afternoon; in an urban park at dusk; in a city centre alone at 11.30 p.m.

Ask 'why?' and keep asking 'why?' to each of your answers.

Related to the above arguments, feminist cultural geographers have argued that 'places' are encoded with messages about who belongs. These messages are read and understood by people and can render individuals quite literally 'out of place'. Places and spaces are gendered. Think about a place that you have been where you have felt uncomfortable: was this in any part due to your gender? An obvious example is the belief common in many societies (not least amongst judges and juries) that women are asking for trouble if they enter public space alone after dark. In this way, the right to public space is denied to lone women after dark (see Valentine 1989; and contrast Mehta and Bondi 1999). Some feminists would argue that this is part of the masculine control of women and space.

The appropriation and control of public space by individuals and groups in power, to make others feel out of place, is a theme that feminist cultural geographers have engaged with in respect of areas of new urban development. The current trend in urban design towards the spectacular or hyperreal has seen the creation of fantasy spaces that are rich in iconography, messages of belonging or exclusion (Harvey 1989b; Knox 1993). Cultural geographers have worked to decode these messages (and those of everyday environments) and to show that power, control and surveillance are very much in operation (e.g. Hopkins 1990; Goss 1993a; Jackson 1998). Others have shown that places and spaces are very deliberately gendered (Monk 1992; Winchester 1992). Feminists also generally accept that an individual's 'sense' and experience of place will vary according to gender and other axes of identity position (Massey 1991; Valentine 1989; 1993).

SEXING SPACE

The ability to appropriate and dominate places and hence influence the use of space by other groups is not only the product of gender; heterosexuality is also powerfully expressed in space.

(Valentine 1993 p. 395)

If we accept that gendered power relations are bound up in and expressed through spaces and places, it is logical that other axes of social division are also made concrete in space. Gill Valentine (and others, e.g. Namaste 1996; Kirby and Hay 1997) has shown feminist geographers that they need to consider that sexuality also exists as a process of power relations in space. In her paper '(Hetero)sexing space' (1993), she argues that most people are oblivious to the way that heterosexual hegemony operates as a process of power relations in all spaces. She comments on the common view amongst heterosexuals that they will tolerate same-sex partnerships if they keep outward signs of their relationships (affection, sexual behaviour) behind closed doors (in the private sphere). She argues that this serves to marginalize, exclude and make uncomfortable 'public' space for lesbian and gay couples, which serves to maintain the heterosexuality of public space. She draws on the experiences of lesbian interviewees to examine the ways that public displays of sexuality are policed through threats, verbal abuse and even physical abuse. Through these power relations and unspoken codes of behaviour, lesbians are forced to find their own safe spaces and are rendered invisible. Valentine discusses different sites as 'heterosexualized spaces': housing, the 'home' and family, workplaces, hotels and restaurants, public outdoor space (such as the street, shopping areas, beaches, public parks).

Many feminists have drawn upon 'queer theory' in their work.

Like feminist theory, queer theory is a diverse body of literature and ideas with multiple roots. It is frequently seen as having emerged out of a literary critisism tradition within the field of gay and lesbian studies ... Now however, it has spread beyond the humanities to engagements with the social and even, at times, physical sciences, where it has continued to evolve and mutate. Queer theory still tends very much to employ *deconstructivist* methods of analysis and to be informed by postmodern and/or poststructuralist modes of thought ... Perhaps most importantly, this includes a political commitment to destabilising binarisms and other oppressive categories of thought.

(McDowell and Sharp 1999 p. 225)

MASCULINISM AND GEOGRAPHY

Gillian Rose published *Feminism and Geography* in 1993, which critiqued the discipline of geography for its inherent masculinism, its phallocentric language (and other form of representation) and its masculinist gaze. She critiqued the subject matter of geography, its underlying philosophical premises (epistemology/ontology), its methods, and its modes of representation. All of these, she felt, contributed to a subject which is less interesting and less insightful than is possible or desirable. Equally she argued, masculinism and phallocentrism alienate and exclude women from the discipline. In terms of the spaces of the practice of geography, she saw much fieldwork with its masculine 'gaze' as disempowering to women students.

TASK

> - What has been your experience of fieldwork since you have been at university?
> - Have you felt in any way that your experience and reaction to this activity are a function of your gender?

Taken to its basics, the feminist critique of science (which is the underlying foundation of Rose's critique of geography) recognizes first and foremost that science is an institution, a social construction, a way of viewing the world and a method of exploring the world. All of these have been taken for granted until very recently. In order to question the validity of science and its method, feminists have deconstructed it (taken it apart to see what it is made of). This is more difficult than it sounds. Brought up in a Western scientific tradition, Western individuals have an in-built understanding and acceptance of the rules and methods of science. Good science is thought to be *value free, objective, replicable, rational*. Feminists have questioned all of these and shown them to be based upon false views of the world and our abilities within it. As England stated, we 'do not parachute into the field with empty heads and a few pencils or a tape-recorder in our pockets ready to record the "facts" ' (1994 p. 84). Feminists have questioned science's assumptions about what can be known and how we can know what we know. Take 'objectivity' for example. Science believes itself to be objective. It thus believes that the scientist/researcher can be objective, that she/he can transcend her or his own self and thus speak and interpret for all. Scientific method requires this. But who are the scientists? White, middle-class, Western men, who pose the questions, choose the methods and interpret the evidence. Can they really speak and interpret for us all? And could they really be value free? Feminists argue that the very language and conceptual frameworks that science uses are value laden (Rose 1993 argues this case in geography). Instead, feminists insist that we need to recognize that we are all subjective beings. Haraway says that as an alternative, 'Feminist objectivity is about limited location and situated knowledge' (1991 p. 190).

In their critique of science, feminists similarly critique claims to universalism and the use of grand narratives (see Chapters 4, 5 and 7). No theories can be seen to hold universal truths for all people. Instead, feminists claim that knowledge is context-bound and partial rather than detached and universal. Feminists have also critiqued science's reliance on dualistic thinking, its essentialism, its masculinism, its spheres of exclusion and its patriarchy (unequal power relations between men and women).

DUALISTIC THINKING

Scientific thought and geography tend to structure understandings around binary opposites, where one is taken as the norm and the other as an opposite, inferior 'other'. Cixous (1981) indicated the gendered nature of this structure.

Where is she?

Activity/passivity,
Sun/Moon,
Culture/Nature,
Day/Night,

Father/Mother,
Head/heart,
Intelligible/sensitive,
Logos/Pathos.

Form, convex, step, advance, seed, progress.
Matter, concave, ground – which supports the step, receptacle.

Man
———
Woman.

(Cixous 1981 p. 90, cited in Rose 1993)

In scientific dialogue, when people are talked about, they are assumed to be 'man', white, straight, Western, middle class unless otherwise stated, e.g. in a recent statement on the BBC evening news, 'many people were killed, including women and children'. Each side of the dualism is defined as the opposite of the other (are men really the 'opposite' of women?), so that if one is seen as intelligent or strong, the other is understood as stupid or weak. Women are seen as everything that men are not. This way of thinking has an inherent hierarchical relationship, not an equal relationship between the pairs. It leads to underlying assumptions in scientific thought about power relations. As Rose discusses, 'What these lists of oppositions represent is a relationship between *A* and *not-A*. This is a field which excludes because it is structured around *A*, the masculine. It is phallocentric. There is no *B*, or *C*, or *D*, *E* or *X*' (1993 p. 75). Such thinking leads to an uncritical acceptance of the 'natural' order of power relations; it leads to the assumption that there is a great distance between opposites, and that differences exist on other related criteria too. Feminist, poststructuralist and queer theory critiques have uncovered these assumptions.

Feminists have argued that the qualities of rationality, objectivity, value freedom, transcendence that science purports are those qualities that women are denied through structures of dualistic thinking. Rationality, objectivity etc. are seen as masculine qualities; hence feminists have claimed that science (and geography) is masculinist. By claiming this, feminists are not agreeing that women are the opposite of these qualitites, but they acknowledge that men have defined them as *being* the opposite. One outcome of this is that women are perceived by society as being unable to be good scientists, as 'woman scientist' appears to be a contradiction in terms (Harding 1991). But more than this, men's science has proven to be less convincing and interesting to the growing number of women researchers and scientists. Rose shows that 'various forms of white, bourgeois, heterosexual masculinity have structured the way in which geography as a discipline claims to know space, place and landscape' (1993 p. 137). In one chapter, she concentrates on geographers' well-accepted method of 'looking' at and interpreting landscape. She argues that 'landscape is a form of representation and not an empirical object' (1993 p. 89). It is not possible to interpret landscape objectively, and interpretations made by geographers reinforce dominant world views or power relations. Feminist cultural geographers have attempted to deconstruct this masculinst landscape tradition in geography. In the final paragraph of her book, Rose calls for a more self-aware and less masculinist geography: 'I want to end by asking for a geography that acknowledges that the grounds of its knowledge are unstable, shifting, uncertain and, above all, contested. Space itself – and landscape and place likewise – far from being firm foundations for disciplinary expertise and power, are insecure, precarious and fluctuating' (1993 p. 160).

CONCLUSIONS

Feminist geography, like its counterpart 'voices from the margins', post-structuralist and postcolonial geography, has drawn attention to the biases and silences inherent in the discipline of geography, and has proposed a new way of 'doing' geography (see Chapter 16) as well as a broadened agenda for what should be studied in geography. Slowly, these messages from the margins of the discipline are being heard. However, it is too early to claim a central position. It is still too often the case that a single feminist chapter is included in geography textbooks as a token. It is still too often the case that feminist geography sessions at major international conferences are overwhelmingly attended by women and few men. More encouragingly, in cultural geography, the feminist message is heard and embraced much more fully. It is also encouraging that much of the new cultural geography has been produced by feminist women and men.

GLOSSARY

Essentialism With respect to gender, this term refers to the tendency in conventional wisdom to attribute gender differences to some natural, biological difference between men and women. Through this way of thinking, all people in a particular social category are regarded as the same. The belief that women are good carers is essentialist, as it assumes that there is some natural quality of all women that renders them good at caring for others.

Gender Traditionally feminists have distinguished 'gender' from 'sex' by arguing that gender is a social constuction, and that its characteristics of masculinity and femininity are also socially constructed. More recently, some feminists have argued that 'sex' too is a social construction, and that gender is not distinguishable from sex.

Masculinism This term has been used to refer to the male dominance of the discipline of geography (Rose 1993) – dominance in terms of subject matter, philosophy and methodological approach (critique of 'objective' science) as well as the academy.

Social construction In feminism, this term refers to the way that society creates women and men, masculinity and femininity. Feminists have argued that there is nothing biological or natural about these categories.

Universalism Refers to claims that are universal in scope. Positivism is criticized by feminists for its universalist claims about the knowledge it creates. Feminists tend away from universalism towards particularism and situated knowledges.

Woman A highly contested term. Feminists often use the term in inverted commas to show that there is no natural or essential category 'woman', but instead, that women are different and are positioned differently depending on other identity characteristics (such as age, race or class).

FURTHER READING

McDowell, L. and Sharp, J. (eds) (1997) *Space, Gender, Knowledge: Feminist Readings*. London: Arnold.

Rose, G. (1993) *Feminism and Geography: the Limits of Geographical Knowledge*. Cambridge: Polity Press.

Women and Geography Study Group (1997) *Feminist Geographies: Explorations in Diversity and Difference*. Institute of British Geographers. Harlow: Addison Wesley Longman.

7 Postcolonial geographies

Pamela Shurmer-Smith

This book is written from Britain, a country which has been constituted out of its imperialist past. It is impossible for people who regard themselves as British to sidestep this imperialism, whatever their colour or ancestry; it is similarly impossible for the formerly colonized and the white settler countries. This is not just a painful history, but one which bequeathed a painful legacy of social inequities and, seemingly indelible anger, shame, fascination, bigotry, resentment, fear and ignorance. Arguably any work that a cultural geographer carries out in, or from, an English-speaking country should be categorized 'postcolonial', but the term is generally reserved for work done with an *awareness* of the outcomes of imperialism. As a consequence, regardless of internal debate, all postcolonial cultural geography has a broadly similar, anti-imperialist, perspective.

As with the geography of gender, it is necessary that the position of any commentator be made explicit in terms of the categories that imperialism rendered important, even though we may hope that one day in the future these categories, based on descent and physical characteristics, will become irrelevant. I write from the position of a middle-aged, middle-class white English-born woman with a working-class background. My ancestors derived few material benefits from Britain's imperial history, but they acquired, and passed on, a world view which placed Britain at the centre. My present middle-class status was achieved as a consequence of the racially segregated education I received when my family migrated to Central Africa, but my experience of the last throes of empire was unusual because my father joined a banned African nationalist party and my allegiances were crossed. I then attended one of the very few multiracial universities in Africa where I studied social anthropology, in that context regarded as radical and anti-colonial, but which has been castigated as 'the handmaid of imperialism' (Gough 1968). I can erase none of these conflicting and confusing facts and they are with me as I write this chapter.

- How do you think of the British Empire: a lost golden age, a violent intrusion, the spread of civilization, a shameful episode?

- Can you identify how your opinions were formed?
- What do you know of your family's place in this history?
- Is there any relationship between your views and your history?

Keep this in mind as you read and evaluate postcolonial literatures.

As good histories of geography demonstrate, much early academic geography can be regarded as an overtly imperialist endeavour (Livingstone 1992). The explorer as a heroic figure glamorized and popularized the discipline. To this day, to step into the premises of the Royal Geographical Society is to be intimidated by the relics of high imperialism, the photographs and trophies of the *men* who discovered, recorded, mapped, named, classified, interpreted and, thereby, tamed what was regarded as savage and strange. In reaction to this, it is not surprising that many contemporary cultural geographers have been swept along by the tide of enthusiasm for postcolonial studies.

Postcolonial theory seeks to uncover the cultural forms that were consequent upon colonialism. The 'post-' refers to a form of critique which gained ground in the light of the demise of colonialism, but the subject matter is not restricted to events which post-date the independence of former colonies. Much work which is thought of as postcolonial retrieves or reanalyses the history of imperialism, destabilizing a Eurocentric viewpoint to look differently at the whole world. Other studies examine present outcomes of imperial relations including those beyond the former colonies – racism, diasporas, flows of information, textual representations, dialogues. Its interests fall predominantly within the realm of cultural politics, as it is concerned with the processes of construction of knowledge, but it is conscious of the importance of political economy.

EUROCENTRISM AND ORIENTALISM

The main thrust of postcolonial studies is to make Eurocentric views of the world problematic; this means destabilizing many of the 'of course' assumptions that have become part of a modern education world-wide. I am painfully aware, as I write my contributions to this book, that too often I look from Britain and couch my explanations in ways which make sense only from that viewpoint.

The rejection of Eurocentrism opens the way to polycentrism and, next, to a more radical claim for decentralization of knowledge. Polycentrism in a context of high interaction generates boundary crossings. Colonialism meant the imposition of boundaries, and decolonization consisted of their appropriation. Postcoloniality ... brackets these boundaries, leaves them behind, and questions cultural nationalism and statist colonization, in the name of multiple identity, travelling theory, migration, diaspora, cultural synthesis and mutation. For instance, popular culture is a hybrid mode that cannot be contained in 'national culture'. Post-structuralism, deconstruction, new historicism and postmodernism are among the theoretical currents informing various reassessments of the postcolonial condition.

(Nederveen Pieterse and Parekh 1995 p. 10)

Postcolonialism has a close alliance with poststructuralism. Edward Said is generally regarded as having founded postcolonial theory and was much influenced by both Foucault's work on the relationship between power and knowledge and Derrida's thoughts on alterity, which he was able to relate to his advocacy for the Palestinian cause. His book *Orientalism* (1978) had a very great impact on Eurocentric complacency. In it he sought to demonstrate that the Orient (as mystic East) was discursively constructed as 'different' out of the imagination of the West. It would be difficult to overemphasize the influence of *Orientalism*, even though it has been criticized for partiality and for marshalling only the evidence which proved Said's argument (Ahmad 1992). It was written for a Western readership and was most successful in embarrassing liberal academics attracted by exoticism into examining their motives and their language. Said's work marked the beginning of the postcolonialist preoccupation with discursive practices, the problem of how power/knowledge is constructed through the use of language. It was also Said who, inspired by Derrida's concept of 'erasure', prompted the interrogation of supposedly 'innocent' texts, most famously Jane Austen's *Mansfield Park*, for their callous acceptance of the structures, and their sidelining of the brutality, of colonialism (Said 1993).

To speak of a scholarly activity as a geographical 'field' is, in the case of Orientalism, fairly revealing since no one is likely to imagine a field symmetrical to it called Occidentalism. Already the special, perhaps even eccentric attitude of Orientalism becomes apparent ... But Orientalism is a field of considerable geographical ambition. And since Orientalists have traditionally occupied themselves with things Oriental (a specialist in Islamic Law, no less than an expert in Chinese dialects or Indian religions, is considered an Orientalist by people who call themselves Orientalists), we must learn to accept enormous, indiscriminate size plus almost infinite subdivision as one of the chief characteristics of Orientalism.

(Said 1978 p. 50)

TASK

> • Find an example of an Orientalist representation in a current news-paper or magazine article or advertisement of Western origin.
> • Think about the clichéd ideas and images of the East which it employs.
> • Where do these originate?
> • Can you relate them to other cultural products you have experienced?

POSTCOLONIAL DISCOURSE

As later chapters demonstrate, archival materials are a valuable source from which we can derive the discourses of colonialism. These can be read not only for what they intentionally transmit, but also for what they reveal of prejudice through the language of normalization of Western concepts in alien settings. The assumptions of both colonizers and colonized seep into the terminology through which they recorded ideas and events, and these can be retrieved from official reports, newspapers, private letters and diaries, political and religious tracts, and even ephemera such as invitations, menu cards and publicity materials. Mary Louis Pratt (1992) has been influential in alerting scholars to the significance of scientific discourses in terms of surveying, classifying and recording as a means of laying claim to territory, and she traces the development of the genre of travel writing as a form of legitimation of imperial attitudes. Her project has been to show how the normalization of a Western construction of alien environments, such as naming places and classifying species and geomorphology according to Western scientific principles, is far from politically neutral.

Archives are always partial and colonial archives even more so than others. They are more useful for unearthing the views of dominant people than those dominated. This does not just mean that colonizers have been more represented than colonized people; it also means that amongst the colonized, the records of the thoughts of educated elites are more likely to be available than those of the mass of people. (It is also easy to form the erroneous impression that virtually all colonials, particularly in India and Africa, were drawn from the upper-middle classes, because educated people not only made the most representations, but their writings were more valued and likely to be retained.)

The influential Subaltern Studies group of historians in India has dedicated itself to retrieving the history of 'subalterns' (a term derived from Gramsci, meaning those below the others). In their case this meant the lower castes and classes whose local resistance to both imperialists and local elites had rarely entered the official histories of independence. Sometimes their evidence comes from reading elite accounts 'against the grain' for traces, other times it comes from oral accounts, including stories, dramas and songs. Gayatri Spivak (1988a), though an admirer of the Subaltern Studies project, wonders, however, whether subalterns can ever 'speak' beyond their own context, given that the conventions of wider communication are denied them and they are condemned always to be represented by others. So far little attention has been paid to the discourses of white colonial subalterns (other than the thriving interest in Australian convicts), but this is an area where Western students could carry out interesting projects which might reveal the roots of contemporary constructions of 'whiteness'.

NATIONALISM

Jameson (1986), writing of what he calls Third World literature, claims that it invariably takes the form of a national allegory, that it writes from a sense of national identity and from a particular nation, rather than having a universal ability to communicate. Although any cultural product anywhere can be used in the cause of nation building, and although conscious fabrication of national culture is a characteristic of independence movements, it is quite startling to contemplate Jameson's sweeping assertion that 'Third World literature' cannot escape functioning as an allegory of the nation, denying it inclusion in world literature.

> In place of the old wants, satisfied by the production of the country, we find new wants, requiring for their satisfaction the products of distant lands and climes. In place of the old local and national seclusion and self-sufficiency, we have intercourse in every direction, universal interdependence of nations. And as in material, so also in intellectual production. The intellectual creations of individual nations become common property. National one-sidedness and narrow-mindedness become more and more impossible, and from the numerous national and local literatures there arises a world literature.
>
> (Marx and Engels 1960 p. 18)

Aijaz Ahmad (1992), whom I rank above all other postcolonial theorists, reacts against the assumption that there is a single thing that can be analysed as 'Third World literature', since such an entity can exist only by dismissing the multiple local linguistic and communicative contexts in which literatures are generated in former colonies. He resents the assertion that any literature can be seen *only* as a reaction to imperialism, or what he refers to as 'merely the *object* of history . . . Politically, we are Calibans all. Formally, we are fated to be in the poststructuralist world of Repetition with Difference; the same allegory, the nationalist one, rewritten, over and over again, until the end of time: "all third-world texts are necessarily . . ." ' (1992 p.102).

> Because I am a Marxist, I had always thought of us, Jameson and myself, as birds of a feather, even though we never quite flocked together. But then, when I was on the third page of this text (specifically, on the sentence starting with 'All Third World texts are necessarily . . .' etc.), I realized that what was being theorized was, among many other things, myself. Now, I was born in India and I write poetry in Urdu, a language not commonly understood among US intellectuals. So I said to myself: '*All*? . . . *necessarily?*' It felt odd. Matters became much more curious, however, for the further I read, the more I realized, with no little chagrin, that the man whom I had for so long, so affectionately, albeit from a physical distance, taken as a comrade was, in his own opinion, my civilizational Other. It was not a good feeling . . . I hold that this term, 'the Third World', is even in its most telling deployments, a polemical one, with no theoretical status whatsoever.
>
> (Ahmad 1992 p. 96)

LANGUAGE, MIMESIS AND HYBRIDITY

It is difficult to overemphasize the importance of the legacy of education in European languages for native elites in colonized countries. This education not only separated elites from masses in terms of access to the dominant language, but it also inculcated alien ways of conceptualizing. Post-Enlightenment logic and scientific thought allied Western educated elites to colonizers and distanced them from their compatriots. This isolation of knowledges has not been diminished by independence and, to this day, elites accumulate global education through European languages whilst masses have local knowledge in local language. This is admirably described in Sanjay Srivastava's (1998) account of an exclusive public school in India. Radhika Parameswaran (1999) has carried out an interesting study of the role of the popular romances published by Mills and Boon in strengthening upper-middle-class Indian women's command not only of English but also of Englishness.

> We have to educate a people who cannot at present be educated by means of their mother-tongue. We must teach them some foreign language. The

claims of our language it is hardly necessary to recapitulate. It stands pre-eminent even among the languages of the West . . . The question now before us is simply whether, when it is in our power to teach this language, we shall teach languages in which, by universal confession, there are no books on any subject which deserve to be compared to our own . . . It is impossible for us, with our limited means, to attempt to educate the body of the people. We must do our best to form a class who may be the interpreters between us and the millions whom we govern; a class of persons, Indian in blood and colour, but English in taste, opinions, in morals and in intellect. To that class we may leave it to refine the vernacular dialects of the country.

(Macaulay's 'Minute on Indian Education', in Ashcroft et al. 1995 pp. 428–30)

One interpretation of this is that elites *mimic* more powerful forces, and in adopting alien cultural forms are inauthentic (Bhabha 1984). A more convincing interpretation is that, in situations in which there is conflict over power, wealth and prestige, people will jockey for advantageous positions, using whatever resources are to hand. Another view of mimicry is that it can be subversive, in that it reflects the practices of dominant people in a mocking, rather than admiring, fashion. But yet another perspective on mimesis comes in the form of imperialist derision at what it sees as imperfect and inauthentic attempts to integrate. Much colonial 'humour' turned on the construction of the figure of the 'brown sahib' with his overly English manners, for example, the effete Babu in Kipling's *Kim* who stands in counterpoint to Kim, himself, who is British, manly and able to pass perfectly for Indian.

Homi Bhabha (1985) has popularized the term 'hybridity' to convey the way in which cultures of imperialism are always entangled. Although I find Bhabha's work very valuable, I have never been able to accept his choice of the term 'hybrid', because it draws upon a biological metaphor where distinct species can mate to produce (usually sterile) hybrid offspring, which are different from, but similar to, each parent. Contained in the concept of hybridity is the notion of 'cross breeding'. Bhabha does not mean this, but if we believe that it is justifiable to question discursive practices, it seems legitimate to ask why a word with racist implications is employed at all, especially with all the overload of the *apartheid* theories of miscegenation. Young (1995) provides a useful perspective on the connection between cultural and racial thinking about hybridization. But, whether or not one likes the term, it is in current usage and there is certainly a need to be able to describe situations of simultaneous separation, convergence and generation.

Q. So if you have to describe yourself how would you describe yourself?
A. Well, I see part of me is white and part of me is black.
Q. But if you were at school what would you say . . . what nationality would you give yourself?
A. English.
Q. English, okay, would you call yourself black English then?
A. Yeah.
Q. And what about your friend Bhavesh – tell me about his background.
A. Well, he comes from Kenya, but he moved over to England and now they live here, but he is . . . he's Kenyan.
Q. And would you call him black English or Asian English?
A. Asian English . . . because I think he's more . . . because you call Asian people Asian, black people black, I would probably call him an Asian English.
Q. And where would his Englishness come from?
A. Well, from living here . . . and he's been to school here and things.

(Hamilton-Clark and Lewis 1998 p. 160)
Hamilton-Clark was a 12-year-old boy, Lewis was his aunt.

An issue of the journal *Postcolonial Studies* (vol. 3 no. 1 2000), which focused on postcolonial cultural constructions of horror, offers some unusual perspectives, drawing upon the way in which horror stories, films and comics have long used the trope of hybridity to convey unease. Universally, a monster is a hybrid being and a haunted space is indeterminate and liminal. As Mary Douglas (1966) has demonstrated, things that are monstrous are associated with concepts of pollution and moral disorder. The various contributors show how postcolonial cultural (con)fusion and loss of a secure sense of locatedness can slide into stories about hauntings and restless spirits (Pettman 2000; Gelder 2000a; 2000b). Such strong cultural devices help fix fears about crossing boundaries and blurring structures, never more clearly than in Clara Law's film *Floating Life* (1996), analysed by Pettman, in which the geographical unhomeliness of migrants from Hong Kong becomes the source of an uncanniness, resolvable only through reconciliation with ancestral spirits. (Pettman comments that Law is claimed by both Hong Kong and Australian cinema.)

Much work on hybridity focuses on the 'fusions' of popular culture, as with the wave of British enthusiasm for Asian fashion, dance music and foods (Sharma et al. 1996) and this is an appealing area of study for many students, who can examine their own involvement and decide whether they would see this as a mixing of two disparate elements or whether it is just the way things are in today's society.

TASK

> • Think about the benefits and disadvantages of the use of the con-
> cepts of hybridity and mimesis when considering examples of appar-
> ent convergence of cultural systems.
> • How do they relate to notions of appropriateness and appropriation,
> sameness and difference, respect and ridicule?

Cultural products are generated for both global and local consumption, and it is interesting to contemplate the difference between those films and novels that one would categorize as 'foreign' and those one would think of as 'international'. The foreign is located in a cultural system which needs translating, but, while the international might originate in the same place, it uses widely comprehensible communicative devices (Clifford 1997). It is recognized in India that there are novelists who write with one eye on the international prizes and markets. These not only write in English (many locally focused authors do this too) but also write of an India which is exotic but rarely appears incomprehensibly foreign to an educated international readership. Arundhati Roy's novel *The God of Small Things* (1997), for example, was translated from English into the major European languages and won international acclaim before it appeared in any Indian language or on the Indian English-language best-seller lists. These transnational novels constitute a genre which could usefully be explored by students of cultural geography anywhere in the world, but their local reception can only be apprehended with some degree of insider knowledge.

There are also the cultural artifacts which are produced with a diasporic audience in mind. A third of the box-office takings for Indian movies is generated from outside the country and this has led to a change in the style and subject matter of commercial films produced in India. On the one hand they are catering to the preferences of the Indian diaspora, on the other they are responding to the aspirations of many people who have not left. Such films may be about modern cosmopolitans and contain scenes involving travel abroad, but the cosmopolitans have securely 'traditional' values and the exotic (European and American) locations are familiar tourist clichés. With the awareness of the overseas market, the boom in Indian middle-class foreign tourism and the increasing globalization of the Indian economy, there has been a noticeable decline in the number of Indian films concentrating on 'boring' social and political issues and an increase in light entertainment about affluent middle-class families. These films can be regarded as nostalgic reminders of home, vicarious enjoyment of a global lifestyle, politically conservative celebrations of 'traditional' family values, endorsement of the market economy, depending upon who is viewing from where (Uberoi 1998; Shurmer-Smith 2000b).

Postcolonial cultural products also emerge outside the former colonies, examining the legacy of imperialism in various contexts. One might well argue that the concept of the postcolonial has been generated to embrace a diverse range of cultural expressions simply because the world has not yet shed the imprint of imperialism. The very notion of postcolonial culture outside former colonies is itself a troubling topic, worthy of study by cultural geographers whatever their personal background. To echo the concerns expressed by Ahmad, why should the place of birth of an author (or his/her ancestors) constitute a significant indicator of authorial difference, cutting across genres?

One should not lose sight of the fact that the many and varied forms of postcolonial cultural expression are responses to complex political and economic confrontations and issues of identity. Racial and ethnic prejudice and discrimination are very real products of imperialism, as are thoughtlessness and cultural arrogance. The task of postcolonial studies is to reveal and confront these in their many disguises. It is rarely a comfortable experience, despite the considerable pleasure and enlightenment that can be derived from the deconstruction of films, novels, poems, music, theatre, autobiographies, travel writing etc. in the light of postcolonial theory. There has been an accusation that, whilst cultural products can originate in any context, the cultural critique (what Mitchell 1992 calls 'post-imperial criticism') which renders this work globally 'acceptable' continues to emerge from the metropolitan powers and their institutions such as universities and learned journals. This process of legitimization is, itself, a topic worthy of consideration by cultural geographers.

ANTI-DEVELOPMENT

The concerns of postcolonialist theory have permeated the thinking of many people who study what is usually called 'development'. The very word echoed thinking which assumed the superiority of the affluent West and consigned the rest of the world to failure. Indicators of 'development' were all Western generated and the whole project rested on benevolent 'advice' from the 'developed' world. People in the West who had made their careers in this field were understandably reluctant to discard a lifetime's work when they began to realize that they had been perpetuating the very structures they abhorred, and development studies reached the theoretical 'impasse' (damned if you do, damned if you don't) charted by Schuurman (1993). Those who were most conscious of their unease started to look at the very ways in which the concept of 'development' had itself developed into its modern form (Cowen and Shenton 1996) and to chart the politics of this idea, notably in the collection edited by Crush (1995). Hutnyk (1996) skilfully destabilized the ethics of charitable volunteering, showing it up as an Orientalist practice. Thousands of NGOs and popular movements the world over were rejecting the idea of 'development' in favour of localized 'anti-development' ways towards a better life.

Aijaz Ahmad (1995) is of the opinion that the whole postcolonialist debate emanates from the concerns of the West and from the moral doubts of people who have migrated to the universities of the West from poorer locations. He feels that in concentrating on the culturalist aspects of the legacy of imperialism, earlier debates about exploitation are being eclipsed. He may just have a point.

GLOSSARY

Diaspora   A population spread out from what they regard as their homeland.

Eurocentrism   Looking at the world from the perspective of Europe.

Hybridity   The product of the fusion of two or more cultures.

Orientalism   A Western way of looking at the non-West which renders it exotic and incomprehensible.

Subaltern   Relating to people without power.

FURTHER READING

Ashcroft, B., Griffiths, G. and Tiffin, H. (eds) (1995) *The Post-Colonial Studies Reader*. London: Routledge.
Ashcroft, B., Griffiths, G. and Tiffin, H. (1998) *Key Concepts in Post-Colonial Studies*. London: Routledge.
Gandhi, L. (1999) *Postcolonial Theory: a Critical Introduction*. Oxford: Oxford University Press.
Nederveen Pieterse, J. and Parekh, B. (1995) 'Shifting imaginaries: decolonization, internal colonization, postcoloniality', in J. Nederveen Pieterse and B. Parekh (eds), *The Decolonization of Imagination: Culture, Knowledge and Power*. London: Zed. pp. 1–19.
Pearson, K., Parry, B. and Squires, J. (eds) (1997) *Cultural Readings of Imperialism: Edward Said and the Gravity of History*. London: Lawrence and Wishart.
Said, E. (1978) *Orientalism*. London: Routledge and Kegan Paul.
Said, E. (1993) *Culture and Imperialism*. London. Vintage.

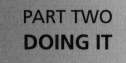

PART TWO
DOING IT

8 Selecting topics for study

*Katy Bennett, Carol Ekinsmyth and
Pamela Shurmer-Smith*

Clifford Geertz points out in *Local Knowledge* that, given the huge variety of possibilities of human behaviour, it is unrealistic for a student of society to try to demonstrate expert knowledge of all possible scraps of belief and practice: 'It is, rather, that one welds the processes of self-knowledge, self-perception, self-understanding to those of other-knowledge, other-perception, other-understanding' (1983 p. 181). When doing active work in cultural geography, it is worth considering whether the 'welding' Geertz recommends is taking place or whether 'others' are mere objects of study, like psychologists' rats. Unless one is careful, work based on discovering and revealing cultural practice is no better than eavesdropping and gossiping.

If this consciousness of mutual 'welding' underlies one's work, selecting a topic of study automatically ceases to be a mechanical process of whittling down a list of potential objects but becomes a search for understanding and empathy. To do this it is best to start with oneself and then work outwards to discover why one is drawn to particular topics. This does not mean self-absorbed studies of 'me'; rather, it means an awareness of the point from which one views the world and a consciousness of why certain things then stand out as worthy of study.

SERIOUSNESS

It is quite common for newspapers in the 'silly season' to trivialize publicly funded research projects by stressing their lack of relevance (no matter that the sums concerned are generally less than a tabloid pays for a juicy story about someone's private life). Every researcher knows how hard it is to secure funding, and floating 'trivial' and 'irrelevant' topics is not the key to success. People who ridicule cultural work operate by tearing research topics out of context; good cultural researchers operate the other way about, by starting with the context and working down to a small and manageable topic which acts as a lens through which to seek a clearer view. Meaningful cultural study is not the study *of* topics but the study of wider issues *by means of* them. No one would doubt that the tourist trade is of considerable economic significance in the Mediterranean region or that it has a considerable social impact in local communities; however, there was a wave of trivialization starting in the tabloid press and working all the way up to a question in the House of Commons about the funding of an anthropologist's work on tourism on a Greek island. The implication was that he was trying to have a long holiday at the taxpayer's expense; the fact was that this was just one facet of a lifetime of work on economy and society in the region.

When required to set up one's own work, the task, then, is to generate topics which can bear the weight of wider issues. If work is useful in testing or refining ideas, one should not be deterred by outsiders who think that the specific subject matter is lacking in gravity. (And remember to extend similar courtesy when next you encounter a physical geographer dropping simulated rain on simulated soil particles. It's just another discursive system.) Topics, in any case, are defined in seriousness according to the values of people in powerful positions and, fortunately, these change. For example, until the 1970s and radical changes in the legal position of women, research on housework or child rearing barely existed, since a masculine establishment regarded the domestic realm as both private and too trivial for analysis. (Young male geographers at the time had no difficulty in impressing the academy with the importance of hops and grapes.) When people deride research topics it probably says more about the perceptions of those in dominant positions in the power structure than about anything else. Today it seems almost incredible that a concern with the moral economy of families should ever have seemed irrelevant, or that Linda McDowell's work on gender division of urban space could seem so revolutionary two decades ago. Similarly, in the early 1990s there were struggles for the body and sexuality to be accepted as legitimate fields of study for geographers. It seems so long ago now, but one might like to generate a list of things which most people would think unworthy of the attention of academic geographers and then to ask the question 'Why?' However, anyone selecting an unconventional topic for an assessed piece of work is advised to make sure that the theoretical justification is unassailable.

OUR OWN RESEARCH

The three following statements were generated by the three of us to relate to our choice of PhD topic. These studies span three decades and are arranged in chronological order.

Pamela Shurmer-Smith

In the early 1970s there was still little work on feminist subjects and what existed focused either on the roles and statuses of wives and mothers or on women's working lives, that is, on women in relation to men. As a single woman, I realized that I lived in a theoretical black hole. I decided to study women outside family, domestic and employment situations in order to try to understand how middle-class English women who were not wives etc. were constructed. I embarked on four years of participant observation in a club dominated by elderly spinsters. The political, social, moral and aesthetic values of the women I studied were very different from my own; in learning how to be with them I gained understanding not just of one small club but of the construction of conservative ideology in wider society. The study fostered an interest in formal processes, particularly bureaucracy, and I continue to be fascinated by the ways in which right-wing morality is constituted. These issues eventually took precedence over my initial feminist concerns and are central to my current work in India.

Carol Ekinsmyth

As an undergraduate in the early 1980s, I had become excited by the field of behavioural geography. The positivist empiricism of much of the published geography of the 1970s had left me cold, and at that time behavioural geography at least appeared to place importance on the cognitive (psychological) processes of individuals, recognizing that an individual's behaviour was never that of 'rational' actor but instead was, in some part, some function of the characteristics of that individual. For me at the time, this came some way to bringing the 'human' into human geography. I became interested in cognitive processes, especially environmental learning, and the ways in which this affected people's behaviour. I decided to conduct research on the way that people learn about places that are new to them. The methods available to geographers to elicit knowledge and to link this to cognitive processes were (and still are) limited; drawing upon developmental theory from psychology, I decided to experiment and devise new methods. These efforts met with some success but were inevitably hampered by the impossibility of the task (the present, far more sophisticated critiques of positivism would undoubtedly deter students from embarking on such a study today – although there are many projects still undertaken in North America in the field). It was easy to show that people differed in their environmental knowledges and their rate of acquisition of knowledge, but far more difficult to link this to their subsequent behaviour. Instead, I had to be content with the general findings that those with the least knowledge of their new environments were the most environmentally timid (were less prepared to explore, travel alone and go to places and were less confident of their ability to navigate in unknown areas), and I was able to suggest some of the likely intellectual, personal and cognitive qualities that distinguished people in this respect. I suggested that 'environmental ability' affected life chances and that children should be educated to develop their abilities. This was the level of conclusion to which the field of behavioural geography was doomed to remain, and this too was part of my conclusion. Psychological factors are still important in geography, especially cultural geography, but they are treated in a quite different way now. In this different way, they are still central to my work on the construction of work and gendered identities.

Katy Bennett

My research, carried out in the mid 1990s, was motivated by feminist debates on theories of patriarchy which I attempted to make sense of in relation to farmers' wives and their families. In brief, some of the ideas that I was reading regarding women's lack of power within the structures of patriarchy did not always make sense regarding my personal experiences of farmers' wives. My mother and grandmother were certainly not powerless within our farming family, although, true to notions of patriarchy, ownership and management of the farm business were in the hands of their husbands and sons. My research on farming families in Dorset attempted to make sense of these contradictions regarding the issue of power (and control) as they were played out through kinship relations and consumption practices. Internal dynamics had repercussions for the responses of family farms to processes of rural restructuring. Researching the implications of, and reactions to, structural transformation of particular households and communities is a continuing theme in my research, and I have since worked in former coalfields and plan to build upon my initial PhD study in light of continuing changes in agriculture.

THEORY INTO PRACTICE

The examples that we have given of our own work should have demonstrated that putting theoretical concerns at the basis of one's selection of topic does not necessarily mean starting with grand theory or high levels of abstraction. The most obvious source of theoretical questions is one's academic reading, such as the current journals and the literature generated by courses.

Since nothing is ever perfect, even the authors one admires should generate questions about different perspectives, different locations and times, the things that seem to be omitted, the comparisons that are not made, the conclusions that are derived from the findings. Sometimes an inspirational study will prompt one to see whether it can be replicated in another location; at other times something will be so annoying that one is determined to show up the author's lack of understanding. Starting from a theoretical question means that one will generate a study topic within an academic context. This means that there is minimal risk of accusations of irrelevance and less chance that it will seem as if a hobby is being pursued in disguise.

TASK

> - Reread an academic paper which has impressed you.
> - Take note of its major theoretical positions.
> - Generate a new topic which could enter into the discussion.
>
> For example, you might read Neil Smith's *Homeless/Global* (1993) and then see whether his ideas about scaling would apply in a different field which had nothing to do with homelessness.

TOPIC-DRIVEN WORK

As supervisors of student projects, we agree that the people who are the most difficult to help are those who express an interest in a topic but are unable to decide what they find interesting about it. This tends to result in a series of false starts, as they proceed down avenues they are not really convinced by, searching for an approach that will make the study academically respectable. Such people are often overly dependent on supervisors for ideas and then find that they do not really think in the way that they are trying to work. Topic-driven work can also lead to the 'scrapbook' approach, which involves little bits of information about a whole lot of tenuously related things. The topic-first problem sometimes emerges from wanting to study too close to oneself – a favourite novelist, a holiday destination, the club one regularly attends. Any of these could usefully be a vehicle of study but, if one does not start from questions which can be regarded as reflecting the concerns of cultural geography, the attempts at analysis of space, place and environment may seem to be merely grafted on to personal enthusiasms, inviting accusations of amateurism. (A frustrating undergraduate dissertation encounter was with a student interested in a disused gravel pit near his home. It took weeks to clarify whether environmental, political, recreational, planning, aesthetic, historical, community, economic or representational issues lay at the bottom of this 'interesting' pit.)

Examples

Tim Cresswell's (1996) study of the subversive use of space and the refusal to know one's place has been an inspiration for many students with a similarly transgressive view of social order. The crux of Cresswell's work is the theorization of the relationship between metaphorical and literal placing, but this is made lively because he constantly works through examples – graffiti artists, new age travellers, gay rights campaigners, antinuclear protesters. If the theoretical position were not so strong, the book would be a collage of episodes; as it is, it is not about the selected topics, it is about the ordering of space as a means of social control. As such, it sensitively draws upon and exemplifies Foucault's philosophy. Cresswell began with his own ideological position and retains the attention of his readers because he cares about the cultural constructions, representations and behaviours he writes about. There can be no better starting point for the selection of a topic for study. Neil Smith's twin concerns with exclusion and gentrification emerge from his desire to work through his Marxist ideology in relation to modern Western cities, and it is his passion and indignation which illuminate his work.

Linda McDowell's work on merchant banks (1995; 1997a; McDowell and Court 1994a; 1994b) arose out of her feminist interest in the gendered nature of occupational segregation and her interest in Iris Marion Young's (1990) ideas about social justice and the politics of difference.

Peter Jackson's research on shopping habits in North London (1999; Holbrook and Jackson 1996a; 1996b) emerged from his interest in cultural politics and the relationship between consumption and identity. His work on advertising and magazines (see Jackson 1991; 1998) springs from similar interests, as well as his interest in the construction of gendered identities.

Gibson-Graham's (1994; 1996) interest in mining town women emerged from a feminist concern that discourses regarding women disadvantaged them by imposing subject positions and identities which maintained their powerlessness. The work was framed in such a way that it not only found out about existing conditions but also tried to help these women develop new, 'liberating' subjectivities (Gibson-Graham 1996).

Claire Dwyer's (1999) work with young Muslim women in West London drew upon her theoretical interests in constructions of femininities and new Muslim identities. She had identified a gap in knowledge and wanted to find out, using ethnographic methods, how the young women viewed their own lives.

TASK

- Without mentioning a particular topic, think of an area of study you are interested in (e.g. patriarchy).
- Think of a broad topic that would allow you to make an interesting contribution to the debates in the field (e.g. structure of rural families).
- Think of a narrower topic that you can research meaningfully (e.g. the particular forms of power relationships which emerge between women of different generations within farming families in rural Dorset).

> • Decide what you would need to do, including where you would need
> to go, what methods you would use, whether you would need
> contacts, what sort of data you need to mobilize (e.g. to Dorset;
> participant observation and interviews; need to contact family to live
> with; data on housekeeping, budgeting, child care, farm ownership,
> inheritance, work on and off the farm).

METHODOLOGY

When generating a topic for study it is important to pay attention to the
sorts of methods which will be required. This is not just a practical question: particular theoretical positions will require particular ways of working.
However, constraints of time, money, personal abilities, political sensitivity
and so on will mean that there will always be some things which, no matter
how interesting, cannot be studied effectively with the resources available.
There will invariably be poor results from work which carries on regardless
when the methods appropriate to the study cannot be applied for practical
reasons. At whatever scale the work is being carried out, one should always
make sure that the appropriate methods and techniques are employed: if
you cannot do what you want, do something else you can do!

Cultural work is not necessarily qualitative or intensive. However, unless
one has considerable funding, and teams of researchers and technical and
clerical assistants, studies which purport to generate new data about
cultural phenomena at a regional or national level are doomed to failure. If
one's interests are at this scale, the next stage after theorization must be to
discover whether there are suitable data sets available to interrogate.

For theoretical reasons, cultural geography tends mainly to be associated
with two major methodological approaches, those of ethnography and
textual analysis. For most cultural geographers, topics need to be selected
which are amenable to these methods; this means that in framing their
topics they need to place emphasis on understanding the particular rather
than the general, the subjective in preference to the objective. (A Portsmouth student once asked women geography lecturers to complete diaries
from which she constructed time-paths. Instead of then interviewing them
to discover the rich personal information which lay behind these, she tried
to generalize, and constructed tables and pie-charts of the activities and
distances. There were only five of us in her survey; one went to India for
fieldwork on one of her two designated diary days and threw the averages
out atrociously. There were good stories about management of time, space,
transport and people to be discovered, but this was just not the way to
discover them. Neither should she have attempted to generalize about five
people.)

WORKING WITH TEXTS

As Chapter 12 explains, the term 'text' has been extended from just words on paper to imply anything that can be interpreted ('read'). It tends to be limited to things which are more or less permanent (e.g. writings, paintings, films, recorded music, landscapes, interior design) or replicable (e.g. theatrical performances, rituals, planned events).

Textual analysis can generate topics for study ranging from the brief classroom exercise all the way up to the full-scale research project, but it is a particularly good way of quickly coming to an understanding of the concept of discourse and the complexities of representation.

I will not cite any, but there have been too many pieces of minor research in cultural geography which provide a single interpretation of a text or complex of texts. Any project in textual analysis should always be designed to unpack the maximum possible meaning, even when this becomes contradictory or confusing. One can start with a poem or a picture and analyse it from within one's own cultural system, but an analysis in the context of the time or place in which it was generated will require more than just brainstorming and may even be impossible to accomplish. It is not really advisable to select a favourite text as a topic of analysis: not only will one be tempted to enthuse rather than analyse, but in pulling out all of the meanings that are not congenial to one's personal reading one is liable to destroy one's affection for the text. (I know someone who declared 'Now you've completely spoiled it for me' when I told him about David Harvey's 1985 reading of Sacré Coeur in Paris.)

Texts can generate topics in comparative study. To take a well-worn example, juxtaposing Fritz Lang's film *Metropolis* (1927) with Ridley Scott's *Blade Runner* (1982) immediately demonstrates a range of devices which have become generic to the communication of ideas about future dystopian urban life (emphasis on verticals and canyon-like streets; association of elites with high spaces and flight, the masses with ground level and below; lack of nature in the form of plants and animals). This could result in the topic 'Spatial techniques in the depiction of dystopia in film'. A straight line of descent can be traced between the two films: viewing some more examples, the topic could become 'A history of the development of spatial techniques in the depiction of dystopia in film'. One might observe that comics and advertisements pick up the same themes: the topic might become 'Intertextual uses of spatial techniques in popular depictions of dystopia'.

TASK

- Pick two more films of the same genre (westerns, thrillers, romances) with at least 40 years between their dates of production.
- Repeat the exercise of generating different topics from them.

Further topics relating to texts can focus on the expression of an ideal or emotion (the nation, freedom, horror, peace, anxiety etc.). Images can be compared and analysed. This is particularly illuminating if different kinds of texts are brought together, such as poems, reproductions of works of art, music, video, advertising materials. There can be considerable similarity in the images across genres, demonstrating the way in which representations work intertextually by drawing upon familiar foundational texts which will have established a fair degree of intersubjective agreement about meaning. Sometimes, however, the task does the very opposite and shows how very different creative artists can be in their materialization of an abstraction. This sort of topic can be handled in a different way, whereby images are shown to people and they are asked to explain what emotions they read from them; here one would be interested in the different subject positions of the people.

These sorts of textual topics are useful as a way of practising cultural geography in themselves and they can vary in scale from a few minutes work to a full-scale project. They can also become the basis of further work in thinking about representation as an activity. Group work on texts can form the basis of group production of representations in folios or exhibitions and performances, including videos and CD-ROMs; it can also be individualized if necessary.

Students often opt to undertake textual analysis for undergraduate dissertation projects. The temptation is to throw one's net too wide. 'Geography and the Romantic poets' would be impossible, but 'Geographical aspects of Wordsworth's poems' would still generate a superficial piece of work. It is necessary to focus more closely (on something like 'Childhood and nature in late Wordsworth'), as this will not only prevent one from wallowing in too many examples, but also allow scope for contextual work.

The most straightforward textual research projects tend to be concerned with the depiction of real places in poems, novels, paintings, films of particular artists (Turner's representations of Portsmouth Harbour, Dickens' London). They can, however, rise far above the level of mere illustration if they assume a theoretical and ideological position, demonstrating *why* the artist took the viewpoint he/she did, explaining how the work contributed to the construction or maintenance of contemporary value systems. Stephen Daniels' (1993) collection of essays on the theme of creation of national identity analyses the work of Gainsborough, Turner, Constable and others to explore their significance in terms of national identity. Like Daniels, Gillian Rose (1993) deconstructs Gainsborough's painting *Mr and Mrs Andrews*, but she does so from the point of view of patriarchy and the equation of women with property. David Harvey (1996) has used the novels of Raymond Williams to enlighten his own moral dilemma concerning the right of academic geographers to engage in local workplace struggles.

As opposed to starting with the oeuvre of a particular artist, texts can usefully be considered generically; here the aim of the research topic is to demonstrate the construction of specialized discourses. Much of this work draws on Foucault, often via Said. Peter Bishop's (1989) monograph on discourses relating to Tibet shows the development of assumptions about remoteness and spirituality which have led to the virtual equation of the real country Tibet with the magical Shangri-La. Laurie Hovell McMillan (1999) continues this theme of the spiritualization and etherealization of the place. There are still plenty of subject areas which have not been considered and plenty more which could usefully be reconsidered. So travel writing as a genre almost invariably serves to distance and exoticize people and places and is a fertile area for study for cultural geographers, but closer to home the sets of novels and films etc. could be mined for information on how values and beliefs are constructed concerning social exclusion, rural life, urban communities, working environments.

The field of postcolonial studies is almost entirely based on textual analysis; here one is attempting to locate texts which embody the ideological issues of imperialism. So far much of the emphasis has been on imperialist texts (Kipling's *Kim* and Forster's *A Passage to India* have been analysed almost to destruction), but there is a rich body of new literature from formerly colonized countries and diasporic communities which can generate new topics in the light of postcolonial theory. Films are rather less easily available outside their countries of origin, but they can be hunted out. Some of the most interesting contemporary postcolonial literature takes on themes of globalization in very sophisticated ways which raise issues of nationalism and internationalism.

Obviously some texts are better than others for exploring representations of space and place. My own opinion is that the most interesting studies emerge from the consideration of texts, whatever their subject matter, where spatial concerns are made overt and where the artist consciously plays with spaces in the construction of mood and narrative. Elizabeth Bronfen's (1999) study of Dorothy Richardson's novel cycle *Pilgrimage* is an excellent account of the way in which a novelist can virtually make spaces into characters; Horner and Zloznik (1990) similarly select space conscious authors for analysis.

Film is perhaps the most spatially aware of all forms of representation. Not only does everything have to be set, located, but space and scale are integral to the mode of communication. The collection edited by David Clarke (1997) is a useful starting point for cultural geographers seeking ideas about the use of film in the collective imagination of place and space, but most of the contributions emphasize character, plot and setting: I keep hoping for a geographical study which prioritizes film's own internal spaces of closeness and distance of shot.

LANDSCAPE AND SENSE OF PLACE

Cultural geographers have long argued that landscape is a text, but they have progressed from the assumption that the task at hand is to produce the authorized expert reading. Arguably, landscapes are best 'read' by groups of people, rather than individuals, and this can form the basis of topics for small exercises in the context of fieldwork. The landscape is never read without awareness of other texts: the influence of years of school geography books is not easily set aside, but neither are memories of previously experienced places, images in films, documentaries, news broadcasts, travel guides, advertisements, all of which transmit their own particular message. Each person will generate multilayered meanings of a landscape; the real interest comes when these complex meanings are put together by a communicating group of people. One then has to confront the often strident multivocality.

These are often referred to as 'sense of place' exercises, as if places had inherent essences to be distilled by trained experts. However, the 'sense' of a place emerges from those insiders and outsiders who exercise their senses, imaginations and memories in relation to it. By definition one's sense of anything is subjective and based on interpretation; this is probably academically interesting only when it becomes intersubjective and senses are communicated. At this point the methods of observation are informed by consciousness of phenomenology; this involves recognizing that people occupying differently constructed social positions experience things differently. Although the experience can be illuminating, sense of place studies are not really capable of producing any great depth of analysis; they are probably best used to generate topics for a fairly introductory sort of activity, rather than promoted into a full-scale dissertation-type project.

ETHNOGRAPHIC WORK

The recognition that meanings are always multiple underpins ethnographic work; one seeks to understand this multiplicity by participating with people in a situation one is interested in, observing and listening to them and perhaps interviewing or setting up focus groups. Ideally, one should end up able to 'see' from every possible position, but the ideal is never realized.

The least traumatic kind of ethnographic work is simple observation. Individually or in groups one can generate topics which will depend upon observing the ways in which different kinds of people use spaces like pedestrian precincts, parks, common-rooms. For example, one can pay attention to which kinds of people seem to dominate and which are dominated by particular spaces, whether groups of people appropriate space differently from individuals, what is regarded as acceptable or as deviant behaviour. If one were planning to take this sort of work further than a topic for a class discussion, such observation would need to be backed up by closer involvement – participation or interviewing to check one's interpretations of the meaning of what had been observed.

PARTICIPANT OBSERVATION

Chapter 13 will look at participant observation. In terms of generating a topic for study, if one is attracted to this method it will be necessary to consider one's own attributes as a part of the selection process. It is not necessary to study groups of people into which one can slide seamlessly; when studying people too much like oneself it is easy to miss what is happening because it seems so ordinary as not to require explanation. However, what is necessary is that one can find a role and carve out a niche. Women can study men; young people can study old; people can study in foreign locations: but it is important always to generate topics which can not only be observed from the position that the participant can occupy, but also be theorized from that position. Attempting a topic which focuses on people one cannot become like, means taking on an adjacent role where one can be involved in interactions. This might mean barmaid in a men's club, helper at a day-centre for elderly people, any job in a foreign country.

In terms of undergraduate work, it is likely that participant observation can be carried out only in the context of a dissertation; it is pretty meaningless to attempt it in the context of coursework (one would probably just be digging for anecdotal evidence in one's own life). Since this sort of ethnographic work can easily flounder in the early stages, it is best if one selects a topic about which one already has a fair bit of knowledge and a setting where there is little risk of rejection. For example, a place where one has worked previously is close enough to know but distant enough to generate topics of research on helpfulness and competitiveness; hierarchical structures; compliance with and subversion of working practices; non-work activities in the workplace; and so on.

INTERVIEWING

If you decide to undertake a project that uses interviewing as the main method, it is useful to consider access to interviewees at the outset. There are plenty of methods texts that outline ways of selecting and recruiting interviewees, but they often make it all sound simpler than it actually is. Sometimes your lack of access to the right people requires you to change the focus of your project; and at other times, your first few interviewees will show you that you need to change the research topic in some way (they may highlight important issues that you hadn't thought of, for example).

They also highlight the advantages of gaining the input of the people who will be involved in your study at the topic-focusing stage of your research. When I (Carol Ekinsmyth) was planning my current research, I started out with the intention of interviewing women and men who worked in the magazine publishing industry (inspired by Linda McDowell's 1997a work on merchant banks, I wanted to investigate the construction of gendered identities and gendered occupational segregation in a cultural industry whose products are highly gendered). After initially finding it difficult to gain access to individuals who worked in the industry, a friend suggested I speak to a friend of hers who used to work for a magazine company. I started the interview by asking the interviewee what it had been like working for her previous company, but found her present position as a freelance journalist far more interesting. This interviewee led me to several other freelances, and before long I had discovered that there was a far more interesting story to be told about the way that the publishing industry uses flexible workers (freelances) to keep its costs to a minimum, how these freelances are often exploited in this position, how gender has a part to play in the organization of freelancing, and how freelances experience this way of working. This in turn has led me to an interest in home as a location of paid professional employment and the social relations surrounding this practice, not only within the sphere of employment, but in the household too. In the process of doing my research the topic itself changed quite radically: this is not at all uncommon and, as long as one carefully monitors the process, this is a valid way of selecting the focus of research (see Ekinsmyth 1999).

CONCLUSION

The topics cultural geographers can choose from are virtually infinite. That is why we did not generate a list from which a selection could be made. It would be meaningless, and might go something like: 'young male street fashion . . . depiction of urban malaise in comics . . . dog ownership and the use of open spaces . . . postmodern architecture in shopping malls . . . West Indian poets in London . . . the arts in place promotion . . . women as buyers and sellers in antique markets . . .'. The two big questions in topic selection are 'Why do I want to know?' and 'Is it possible for me to find out?'

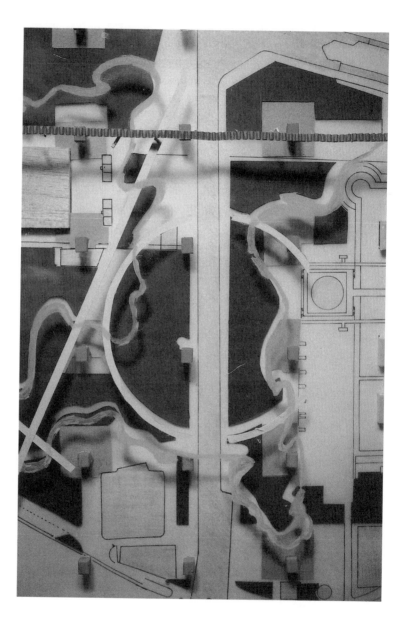

9 Methods and methodology

Pamela Shurmer-Smith

It should be obvious that contemporary cultural geographers are acutely conscious that their task is not a simple matter of studying an out-there reality. The way one goes about defining the area of study, the ideological position one operates from, the way one relates to the subject matter, all contribute not just to one's style of working and representation, but also to the creation of a new cultural product, however small. How one goes about one's work, regardless of whether it is to produce a research publication, a documentary video, an essay or just a contribution to a discussion, becomes a part of what is being studied. This means that a 'recipe book' approach to methods is never going to be effective.

In this part of the book, none of us intends to lay down a set of rules relating to 'good' practice for doing cultural geography. This does not mean that we do not, as individuals, have clear opinions about how to go about our work. We have, at times, argued passionately (and even acrimoniously) with each other about research methods and ways of writing. What one *does* in relation to one's field of study is intimately connected to what one *thinks* about it, but, although theorization is absolutely vital, I believe that, ultimately, the *doing* is more important. Methodology is not just a matter of practicalities and techniques, it is a matter of marrying up theory with practice (the '-ology' of method). When one adopts a particular theoretical position, some methods will suggest themselves and others become inappropriate, for both theoretical and practical reasons. So, for example, if one takes the view that all human beings are unique, with a uniqueness which comes from within, there would not be much point in conducting extensive questionnaire surveys or using 'scientific' methods (which is part of why behaviouralism fell from grace). The method would not be capable of singing in tune with the theory.

When we argue with our colleagues about the virtues of one method over another, we are inevitably disagreeing over more than method; we have to face the fact that methodological disputes are based on different notions of what an investigation is trying to achieve. This can include the questions of what motivates the work, what one believes to be morally acceptable, what is aesthetically pleasing.

My own preference when conducting social and cultural work is for long-term commitment between myself and the people I am working with. I see research with people as requiring that I become part of what I am studying, with the implication that I cannot avoid being politically and morally bound up in it. This means that I cannot contemplate doing research in the manner of a raid – dashing into 'the field' with a Dictaphone and a schedule of interviews, dashing back to the university to transcribe and perform a content analysis. I see this as the theft of other people's words and ideas. However, some people think that if an activity becomes a part of life, the researcher cannot be detached enough to warrant calling it research. Katy Bennett will consider issues relating to participant observation in Chapter 13.

Close involvement in ethnographic research has unavoidable moral implications. In unearthing what seem to be innocuous facts, one can hurt people's feelings or puncture their self-esteem. In some situations, ethnographic research can even uncover knowledge which can be dangerous or incriminating: just promising anonymity is rarely enough to conceal someone's identity. Often conscience directs one not to make intimacies known which would prove one's effectiveness as a researcher to those who like to sit in judgement over these things. I would caution against close and intimate work with people unless you are concerned with building up deep understanding, and are prepared not just to share their lifestyle temporarily but also to share their identity. Scooping juicy case studies seems rather despicable if one genuinely makes a commitment. In such a case, the understanding obtained from sharing in the private matters of a community is invaluable for making sense of cultural practices, such as texts and festivals, which are in the public domain. It is in such situations that one can best act as a representative – not someone who bares all, but someone who knows what to remain silent on.

Despite my distaste for heavy reliance on interviewing in cultural research, it is important to consider when it is a valid method. Interviews are useful for getting people to state the normative values of the community (the way that it is felt things 'ought' to be). For this reason they are particularly useful when dealing with leaders and public figures, who quite often will not talk to a mere researcher in any other way, and even then demand a prepared list of questions in advance. The interview, no matter how skilfully conducted, is artificial; this means that it is valuable for tapping into self-conscious practices, knowledges and beliefs. Interviews are also useful for revealing story-telling techniques, how accounts are made out of experiences and observations. Thus, an interview is less likely to reveal what it was really like when an aged interviewee was young than to reveal how old people in a particular cultural milieu construct and represent their childhoods and, thus, their presents.

For those who are squeamish about working directly with people, textual analysis may appear attractive, but it reveals only the fragments which people have chosen to represent by translating them into symbolic form. Textual work needs, then, to be tackled by people whose theoretical interests lie in terms of a society's processes of knowledge creation, including selection and encoding. The fragments available can include the archives of the past, the official data of the present, artistic and mediated products. They can be rich in meaning, particularly when intertextual comparisons are made, but it would be rash to try to understand a social context solely by reference to the texts it produces. Representation is not just a matter of presenting again (re-presentation), it is also about putting a particular view forward in a privileged fashion; all representations then contain an element of self-censorship even if they do not contain embellishment. Textual analysis (including landscapes, pictures and music) is valuable for understanding the processes of representation and the ways in which these act in dialogue with everyday life, but it is difficult to derive any meaning from a text unless one already has some knowledge of the context in which it was produced.

The aim of research in cultural geography is to shed light on shared values and behaviours relating to the geographer's concerns of space, place and environment. If one accepts that culture is not a thing, one ceases trying to decipher whole cultures and starts to realize that all one can hope to do is to begin to understand the ways in which meaning is constituted in particular situations. If one believes that meaning is constructed by people interacting with each other, participant observation will beckon as a method (or perhaps focus groups) since the researcher will be included in the construction work; if one sees meaning as made up of self-conscious processes, then it will probably be interviews; if one has faith in the power of discourse, the deconstruction of texts and transcripts will probably be appealing. Cultural geographers rarely collect extensive data themselves, although they sometimes gain ideas for intensive research from the questions thrown up by published data sets. The reason for this partial rejection of extensive methods is that, until statistical methods become much more sophisticated, they cannot handle the multiple strands that compose any culturalist's interests in even a simple topic. There is, however, every hope that the development of neural networks will, one day, allow cultural geography to take on new ways of working. (The real problem, of course, lies deeply buried in the traditions of academic culture and realms of cultivated ignorance.) Cultural geographers do, however, regard techniques of enumerating, tabulating and correlating as cultural products, capable of being deconstructed, as in Appadurai's (1996) account of the role of statistics in imposing imperial control in India.

Most of us dabble with different forms of research for different aspects of our work. Often we do this in the name of method triangulation, but most of us put our real trust in just one favoured method which sits most easily with our view of the world, seeing other ways of working as reinforcement. When one is reading, it is important to ask why a researcher has employed a particular method, since this will give vital clues as to the way in which the subject matter has been approached. Sometimes people refer to a method but present no information which would obviously emerge from having followed it, giving clear clues that they were not easy with this way of working but thought that 'good practice' implied that they ought to have adopted a fashionable technique. It then becomes difficult to trust what that person has written.

FURTHER READING

Flowerdew, R. and Martin, D. (eds) (1997) *Methods in Human Geography: a Guide for Students Doing a Research Project*. Harlow: Longman.

10 Extensive methods: using secondary data

Tim Brown

A chapter on extensive methods within a cultural geography textbook would, at first glance, appear slightly out of place. As a close inspection of any one of the textbooks on the subject would reveal (see Shurmer-Smith and Hannam 1994; Crang 1998; Mitchell 2000), this is an area of research that receives very little attention in a predominantly interpretive sub-discipline. Yet, if we consider the role played by extensive data in the production/construction of knowledge about the social and cultural worlds we inhabit, it would appear that such information should be relevant to any student interested in cultural geography. For example, our under-standing of key concepts such as 'social class' and 'consumption' is partially constructed through reference to various sources of social survey data, such as the Census or other surveys that generate ideas about consumption patterns and consumer lifestyles. (One might think here of yuppies and dinkies, defined by their lifestyles or household characteristics: young urban professional people; and double income, no kids.)

Complex interaction between qualitative and quantitative ways of know-ing is rarely reflected in the debate over these two paradigmatic research methodologies. Indeed, for some, the turn away from extensive data has apparently hardened, with individual researchers continuing to position themselves, and by extension their students, on either side of the methodo-logical divide (Philo 1996; Philo et al. 1998). This is a point on which Thrift and Johnston (1993) comment: 'many of those involved in the cultural turn now have little sense of other developments, or simply cannot believe their relevance . . . quantitative work is now often simply dismissed by younger workers with reactions that are so virulent that they can only be about the identity of those concerned' (cited in Phillip 1998 p. 265). However, in spite of this hardening of resolve, the debate over the use of quantitative methods and data has continued, a point that recent editions of the *Professional Geographer* (1995) and *Environment and Planning A* (1998) attest to.

Given this newly found scope for dialogue, there are two central issues that I wish to address in this chapter. The first is the value of extensive data to the production of cultural geography research, asking what they could add to a study that other methods currently do not. The second, and related, issue is the question of how extensive data can be integrated into a largely qualitative tradition, asking whether it is possible to adopt a multi-method approach.

EXTENSIVE METHODS: DEFINITIONS AND DATA

The difference between quantitative and qualitative or, following Harré (1979) and Sayer (1992), extensive and intensive, research paradigms is associated with key underlying principles regarding the production of knowledge. Here issues of epistemology, methods and techniques all emerge as important elements in helping to explain, understand or interpret the object(s) under investigation. In terms of designing a research project, it is arguably the latter that is of most importance. After all, the methodology used in the data collection process influences the type of data collected, and it is here that the researcher's epistemological and theoretical position can be most easily compromised. Therefore, in order to use extensively pro-duced data in cultural geography one must be sensitive to the implications of such differences. One way of thinking through this difference between the two methodologies is to note down their main characteristics and place the corresponding elements alongside each other: for example, large sample size versus small sample size, or nomothetic versus idiographic.

Extensive methods present a particular view of the world. By asking different questions (what are the patterns? what is the frequency?) and by employing different methods and analytical techniques (large-scale surveys with statistical analysis), extensive research produces data that are used to develop causal explanations and test specific hypotheses. As one would expect, there are a number of critiques associated with this particular perspective: the claim of objectivity and an assumed legitimacy; the prob-lem of measurement and definition; the (false) division constructed between researchers and their object of study (see Brannen 1992; Sayer 1992; McLafferty 1995; Phillip 1998). Outside such basic epistemological and ontological disagreements, a central concern is that quantitative data exclude as much as they reveal. As Haraway (1991) suggests, they con-struct a 'false vision' which denies the impact of one's own subjectivity in the production of knowledge.

However, while this view of extensive methods prevails, there is some debate about the necessary link between this approach and the production of positivistic accounts. The question that people like McLafferty (1995), Moss (1995) and Philo et al. (1998) ask is whether a critically aware quantitative methodology is possible. Such an approach would not aim to supplant qualitative research. Rather it would seek to add another research perspective, one that recognizes the socially constructed nature of knowledge and does not seek to construct singular truths or lay false claims regarding its legitimacy over other forms of knowing. There are, then, two points that I'd like to raise here. The first is that extensively produced research material would often be of value to cultural geographers if the numbers were more sensitively deployed. The second is that attention should be paid to the process(es) behind the production of such material.

SOURCES OF EXTENSIVE DATA

The information most likely to be accessed by geographers is derived from either government departments and their related agencies and institutions or non-governmental bodies such as independent research agencies, universities, private companies, the media, charities, religious groups and political parties. The data collected in the UK, as elsewhere in the world, are quite considerable and extend to material regarding the opinions, attitudes, prejudices, values, lifestyles and behaviours of individuals and groups. The aims of this section are, first, to highlight some of the main sources of UK data (given the space available it is not possible to extend beyond the UK, but most countries have broadly equivalent data) and to provide some insight into what sorts of information are available; and, second, to produce a strategic guide for locating the data.

The data sources

An obvious starting point for many human geography research projects is the population Census of Great Britain (http://census.ac.uk). Conducted every 10 years by the Office of Population Censuses and Surveys (OPCS), which has now been incorporated within the Office for National Statistics (ONS), this source provides data on many aspects of the UK population. The information available in the Census, which does vary, can be divided into two sets: the 100 per cent and the 10 per cent questions. The former refers to the basic questions that are processed in their entirety: demographic characteristics, including age, sex, marital status and ethnicity, location, employment status, housing and car ownership. In contrast, the 10 per cent questions are those that are only partially processed; this group in the 1991 Census included relationship in household, occupation, employer and journey to work (see Openshaw 1995).

A key benefit of these data is that elements of them are available at a variety of different scales, from the individual, household, enumeration district and ward to the regional and national. Furthermore, this ability to distinguish between spatial scales has encouraged many local authorities to process and analyse the data for their regions. For example, the Portsmouth City Council published the *Census Results for Portsmouth: Key Facts and*

Figures (1993). However, while the Census might provide useful contextual data for cultural study within geography, it should not be approached uncritically. It should be remembered that the categories and definitions used are not necessarily compatible with interpretive understandings of the social world.

Hakim has defined secondary analysis as 'Any further analysis of an existing data set which presents interpretations . . . additional to, or different from, those presented in the first inquiry as a whole and its main results' (1982 p. 1). A useful example of this is the debate surrounding questions on ethnicity and relationship in household. What we find here is that important issues arise regarding the use and further analysis of Census and other extensively produced data. In the 1981 Census all members of a household were assumed to be of the same ethnicity as the head of household (Fenton 1996), whereas the members themselves would not necessarily agree with this. By the 1991 Census this assumption had gone, but now there were problems relating to the undifferentiated category 'white'.

There are many other sources beyond the Census providing data related to specific areas of interest. While I will simply refer to these sources here, the notion that statistical surveys conceal as much as they reveal should be kept at the forefront of one's mind. One needs always to ask questions about the data, why they were collected, what questions were asked, what assumptions were being made by those who generated the questions. This aside, sources other than the Census come in many different forms. For example, government departments and associated agencies commission surveys that relate to their particular area of administrative control: the *Health Survey for England* (Department of Health), the *Labour Force Survey* (Department of Employment), the *United Kingdom Tourism Survey* (English Tourist Council) and the *Survey of English Housing* (Department of the Environment, Transport and the Regions). In addition to these commissioned surveys, there are the statistics produced by the departments themselves, usually on an annual basis: the Home Office produces data on crime and asylum seekers in its annual bulletins *Recorded Crime Statistics* and *Asylum Statistics* and the Department for Education and Employment provides statistics on schools in its annual *Statistics of Education, Schools in England*.

What many of these statistics produce is a static snapshot of the various aspects of British society with which they are concerned. In order to introduce a temporal element into this snapshot, other surveys and statistical publications produce longitudinal data from studies that track the objects under investigation over time. There is a range of possible sources of such data, including *Social Trends* which focuses on topical social issues; *Regional Trends* which provides a broad based description – demographic, social, industrial and economic – of the regions in the UK; and *Population Trends* which covers population and demographic information. Other sources of longitudinal data are the *General Household Survey* (GHS) which includes data on housing, employment, education, health and social services, transport, population and social security, and the *British Social Attitudes Survey* (BSAS) which, like its US and European counterparts, produces data on social, economic, political and moral issues.

TASK

Locate a source of official numerical data relating to something you are interested in (e.g. ownership of goods, marriage and family, housing tenure, education, health).

- Who commissioned the research?
- Who produced it?
- What do you think it was produced for?
- Could the data be reworked for different purposes?
- Do the data contain the answers to the questions you would wish to ask?

Thus far, I have focused on what might be described as 'official' data. It is important, however, to be aware of other available resources, especially those that fall into the broad category of 'unofficial' data. As indicated in the introduction to this section, there are a vast array of organizations that produce such material. Examples that may be of interest include *Market Intelligence*, a monthly publication produced by the independent research organization Minitel, and the *Lifestyle Pocket Book*, produced by the Advertising Association. Both of these provide statistical information on the products being consumed and also on consumers themselves. For those interested in the media, there is the *National Readership Survey* (http://www.nrs.co.uk) which provides information on the readership of national newspapers and consumer magazines. The research publications produced by the Independent Television Commission (ITC) also provide material on viewing statistics and on specific issues such as nudity, ethnic representation and public opinions.

Sources such as these are generated by a range of independent organizations that are either required to produce statistics, such as the ITC, or do so for commercial reasons. Outside these, other bodies, such as charities, universities, political parties and professional organizations, produce data on specific issues or areas of interest. Taking the example of homelessness, information is produced annually by The Institute of Public Finance in their *Homelessness Statistics* or biennially by Crisis, the UK charity for homeless people, in their *Homelessness Factfiles*. It should also be remembered that public libraries often hold information that is collected about their local area. For example, in Jackson's (1988b) work on 'Neighbourhood change and local politics in Chicago', he employed a range of quantitative material to add context to his study. In addition to relevant Census data he also looked at electoral returns, local community fact books and other such sources to develop a broad understanding of the social, political and demographic characteristics of the area of Chicago in which he was interested.

Locating the data

In terms of locating the data there are a number of strategies that might be employed. As a starting point, there are various written guides including the *Guide to Official Statistics* (Office for National Statistics 2000) and *Sources of Unofficial UK Statistics* (Mort and Wilkins 2000). Then there are the online databases provided by the Office for National Statistics (http://www.statistics.gov.uk), the Data Archive at the University of Essex (http://www.data-archive.ac.uk) and the MIMAS database at the University of Manchester (http://www.mimas.ac.uk/). In the case of the Office for National Statistics, information of potential interest to cultural geographers is located under the general title of 'social surveys'. This division of the ONS conducts social survey research for government departments and has organized the information that it provides into 13 separate theme areas, ranging from 'social and welfare' to 'transport, travel and tourism'. Each theme area is described in full on the ONS website, giving details of the topics covered, the main organizations involved, the publications produced and the statistical sources upon which these publications are based.

With regard to the ESRC sponsored data archive, data are held on over 5000 separate studies which can be explored using the BIRON (Bibliographic Information Retrieval Online) search engine. Using BIRON is reasonably straightforward and allows researchers to access the archive's 38 'major studies' – including the *British Household Panel Survey*, the *General Household Survey*, the *British and Northern Ireland Social Attitudes Surveys* and the *Youth Cohort Study of England and Wales* – or to search for information using a number of subject categories. The value of this process is that it results in detailed descriptions of the available studies and provides a comprehensive bibliography. This is particularly useful because it documents reports that have been produced by the studies' principal investigators and by others using the data for secondary analysis. The archive held by MIMAS offers a similar service to that of the ESRC data archive, with one key difference: the data held are from major cross-sectional and longitudinal surveys only.

When it comes to 'unofficial' data the task becomes much more difficult, although the data archive at Essex is an important resource for identifying studies conducted by university sector researchers. A further strategy for locating data produced by university researchers is to conduct bibliographic searches on either BIDS (http://www.bids.ac.uk) or the Web of Science (http://wos.mimas.ac.uk). While this process may not lead directly to a specific data set, it will help to locate academic work being conducted in a particular area. For research outside this sector, probably the best starting point is Mort and Wilkin's (2000) reference guide which, amongst other things, points to sources of data on advertising, consumption, homelessness, social attitudes and tourism. However, if none of these approaches proves successful, a useful alternative is to focus attention directly on specific charities, interest groups or other non-governmental bodies. This can be done increasingly over the Internet and is an important process because it may provide data that are positioned alternatives to an 'official' line.

REDISCOVERING NUMBERS

On the face of it, the sources of extensive data that I have mentioned in this chapter appear rather arbitrary. This is deliberately so, as the main point that I have sought to get across thus far is that various agencies and institutions record all aspects of our social lives in some way or other. Indeed, it could be argued that such information gives an insight into the ways in which our lives are imagined by government, by policy-makers and by interested others. This diversity of statistical resources should, then, help to make one aware that statistics are not 'social facts' but rather are 'social and political constructions which may be based upon the interests of those who commissioned the research' (May 1997 p. 65). This point was clearly demonstrated when the Thatcher administration sought to halt the progress of *The National Survey of Sexual Attitudes and Lifestyles* at the height of UK concerns about HIV and AIDS (see Wellings et al. 1994). Thus social statistics, of whatever kind, help us to understand the dynamics of society – whether in relation to race, class, ethnicity, gender or age – and to position these understandings in the context(s) in which they were produced.

WHEN IS IT OK TO COUNT? USING EXTENSIVE DATA

While social statistics have much to offer cultural geographers, the question of their use in post-positivistic research is one that has received a great deal of attention in recent times. Such a rethink has come about, in part, because of a growing concern with what Philo (2000) recognizes as the 'desocializing' and 'dematerializing' tendencies of research conducted under the banner of the 'new' cultural geography. While not criticizing the 'preoccupation with immaterial cultural processes', Philo reflects upon cultural geography's tendency to 'focus all too quickly . . . on identity politics and cultural representations' (2000 p. 37). The concern here is that, in the 'doing' of cultural geography, important questions regarding the material and social spaces that surround our everyday selves are being forgotten – questions that might, for example, place the analysis of identity politics or cultural representations within the context of broader social processes relating to poverty, racism, homophobia, sexism and so on. The purpose of this section is to explore ways in which such a (re)connection between quantitative methods and interpretive cultural geography might be made.

A very similar question was directed at feminist geographers in their discussion on quantitative methods published in *Professional Geographer* (1995). In asking 'Should women count?', the authors sought to explore how the main concerns of feminist geography might be further advanced by (re)introducing quantitative data into their research. This did not involve a criticism of qualitative approaches, which were seen as a vital means for reconnecting researchers and empowering the researched, but rather stemmed from a recognition that quantitative data might add other important perspectives. For Moss, such a use of 'counting' might take a particular form: 'In the refined epistemological orientation I propose here, numbers give (re)presentations context – they situate qualitative interpretations within the wider social order and political economy' (1995 p. 447). Importantly, however, this use of quantitative data to provide context is located within what Moss calls a 'refined critical practice'. Here, the notions of reflexivity, embeddedness, situated knowledge and (feminist) objectivity are employed to frame the use of quantitative data.

Many of the issues raised in this debate are increasingly evident in research conducted by social and cultural geographers. One such example is Hubbard's (1999) investigation into the question of social exclusion and marginalization as it is played out amongst female sex workers. The specific aim of the study was to engage female sex workers in meaningful conversations about their experiences of the 'margins'. Clearly, this type of investigation required Hubbard to adopt a critical approach, especially given his own gendered position within it: 'I had to consider whether my attempt to explore the exclusion of sex workers would be anything more than a masculinist attempt to appropriate the feminine "other"' (1999 p. 231). However, while Hubbard was sensitive to his own positionality and the possible implications of it, he came across a number of insurmountable problems. Of most importance was the fact that the sex workers themselves were unwilling to allow Hubbard into their world(s).

For Hubbard, however, the answer was not the wholesale abandonment of the research project but rather the recognition that other forms of research material might be deployed. As he argues, intensive data are not the only sources of 'useful' information for identifying and contesting exclusion. Following Moss (1995), Hubbard saw extensive data as a means to 'identify the general characteristics, lifestyles and forms of oppression experienced by sex workers' (1999 p. 234). There are a number of points that need to be made here. The first is that, as researchers, we must recognize that positionality is sometimes not enough. While we might have the best intentions in mind, it may be that we, ourselves, represent the 'forces of oppression' that we seek to understand. A second and related issue is reflected in Hubbard's own response. That is, when extensive research is positioned within a theoretically informed and critical methodological framework it can help us to better understand broader social processes and group characteristics. It is not a replacement for qualitative research, but neither is it necessarily the antithesis of it.

The aim of this chapter has been to establish the notion that extensive research has a role to play in cultural geography research. This is not an attempt to produce a new canon; rather, in the spirit of plurality and multiplicity, the purpose was to (re)open the 'black box' that surrounds methodological approaches in cultural geography (see Philo et al. 1998). While an apparently simple aim, the job of opening up a dialogue between the two research paradigms is a complicated task. The main reason for this lies in the clear epistemological and ontological differences that exist between the two. However, as I have indicated, extensive data can be of some value to an interpretive sub-discipline such as cultural geography.

On a very general level we should recognize the impact that social surveys and other related statistics have on understandings of the social and cultural world. Concepts such as social class and consumption do not emerge in isolation from such data but are actively constructed through a dialogue with it. Going further, one might argue that the very act of commissioning, of doing, of analysing and of interpreting such data is a cultural act. We might begin to ask why such surveys were commissioned, how they were conducted, what questions were asked and what categories were constructed. In my own area of interest, HIV and AIDS, the act of defining and labelling individuals on the basis of their supposed risk group status had profound implications with regard to the popular and professional understanding of the epidemic. Therefore, at the very least we might begin to examine the processes behind the production of extensive data and the cultural implications of such data.

I have also suggested that an engagement with extensive data should go much further than this. In the examples cited, quantitative data have actually been used as part of a multimethod approach (see Phillip 1998). Whether the data are taken from the Census or from other quantitative sources, they have been used in a number of ways. In the case of Jackson's (1988a) work, the data were used very much in the developmental stages of a more qualitatively oriented project. For Hubbard (1999), quantitative data were used to reveal broad social characteristics and to highlight exclusionary processes that impacted upon a specific social group. What stands out in studies such as these is the fact that the material used was situated within a critical theoretical and epistemological framework. This is of particular importance because, as Hodge notes, methodologies are 'powerful extensions of epistemological and ontological positions' (1995 p. 426). Therefore, while we may wish to be pragmatic in our choice of methodology we must also be sensitive to the implications of the approaches that we employ.

FURTHER READING

Levitas, R. and Guy, W. (1996) 'Introduction', in R. Levitas and W. Guy (eds), *Interpreting Official Statistics*. London: Routledge.

McLafferty, S. (1995) 'Counting for women', *Professional Geographer*, 47 (4): 436–42.

May, T. (1997) *Social Research: Issues, Methods and Process*, 2nd edn. Buckingham: Open University Press.

Phillip, L. (1998) 'Combining quantitative and qualitative approaches to social research in human geography – an impossible mixture?' *Environment and Planning A*, 30: 261–76.

Philo, C., Mitchell, R. and More, A. (1998) 'Reconsidering quantitative geography: the things that count', *Environment and Planning A*, 30: 191–201.

11 Using archives

Kevin Hannam

Historical cultural geography based on archival research has arguably been at the forefront of theoretical advances in geography in recent times. Driver has noted that: 'any division between non-historical human geography, oriented to the present, and an historical geography oriented to the past can no longer be sustained . . . thinking historically . . . is an essential part of doing human geography' (1988a p. 504). Archives are the primary sources of historical information, much of which may be useful in researching previous cultural landscapes and lifestyles. However, archives may also contain important information pertaining to contemporary culture.

The records contained in archives are made and used in accordance with organizational routines, and depend for their intelligibility on shared cultural assumptions. More politically, archives can also be thought of as an integral part of the apparatus of modern government. The formation of archives is a characteristic of modernity that emphasized values of ordered, systematic knowledge and the scientific search for truth and classification. In this context Michel Foucault provided the following forceful argument.

> The idea of accumulating everything, of establishing a sort of general archive, the will to enclose in one place all times, all epochs, all forms, all tastes, the idea of constituting a place of all times that is itself outside of time and inaccessible to its ravages, the project of organising in this a sort of perpetual and indefinite accumulation of time in an immobile place, this whole idea belongs to our modernity.
>
> (Foucault 1986 p. 26)

The formation of archives in the nineteenth century heralded a new relationship between governments and history, whereby the former came to be the major guardian of knowledge about the past. Archives were designed to provide a new setting for history, one that would be open to all and lift the cultural level of the wider population (Bennett 1995).

Records kept in government archives would thus be given a privileged status. Indeed, as Tosh has noted, 'official records have a certain anonymity, which warrants their treatment by members as objective, factual statements rather than as mere personal belief, opinion, or guesswork' (1991 p. 173). A tension would remain between the supposed universality of the knowledge collected and the culturally partial ways in which this knowledge was, and still is, ordered.

Despite their outwardly omniscient character, archives are selective in the sources that they collate. Indeed, they tend to focus on generally male, generally statistical and generally elite sources of knowledge. In the UK, the Durham County Record Office openly admits that 'much of the vast quantity of paper which is created by organisations is not worth keeping permanently and archivists are continually making decisions on what should be preserved' (http://www.durham.gov.uk/recordoffice/). Archives receive much of their material in the form of gifts or long-term loans from people and organizations that regard themselves as significant in some way. However, material is sometimes deposited but withheld from public consultation as it is deemed to be too sensitive. Archivists, in turn, select from this material, documents that they perceive to be significant for the public. In order to be significant the material has to be specific to a particular organization, and it must be authenticated in some way, i.e. handwritten, dated and signed. Many letters are retained only if the content is deemed significant by the archivist – an ultimately subjective decision. Ephemeral or repetitive material is either destroyed or returned to the owners, although often an example is retained. Decisions as to what is given to archives and what is kept are inherently political and reflect notions of what constitutes valuable national or regional heritage. This implies that the history of some people is more valuable than that of others, with the result that elites are over-represented whereas marginalized groups can almost disappear. Over time ideas may change on this score as a new present starts to write its history and former 'rubbish' becomes valuable.

ARCHIVES AS NATIONAL HERITAGE

Belonging to a nation is generally seen as a democratic, shared emotional experience that subsumes other communal differences (Anderson 1990). Archives, monuments, museums, celebrations, schools and the mass media all attempt to convey a historical sense of national identity. However, these are all politically contested forms of remembering as certain social groups have been better able to exclude competing memories or visions of the past and impose their ideas of what constitutes valuable heritage upon other more marginal social groups. Archives often claim to be politically neutral repositories of evidence of national culture and memory; however, they are inevitably far from neutral, as they help to define what a nation means.

In uncritically adopting an 'official' government archive as the primary source of knowledge, a researcher may adopt the view of the established government and ignore, or at the very least treat as secondary, the voices of marginalized people. Indian history, for example, has been associated almost exclusively with the elite achievements of various government personalities, institutions, activities and ideas. Such elite centred histories ignore and thus fail to explain the autonomous domain of the actual, everyday politics of the people (Guha 1988; Richards 1992). However, if we were to ignore official sources altogether, there would be profound difficulties in terms of attempting to gather enough evidence of marginal cultures. Hence, because of the difficulties of retrieving documents, sometimes the cultural identities of marginalized groups can only be read through the sense of crisis they create in official government texts.

> From the first voyages of discovery to the social revolutions of the present day, the story of Canada is best told by the people who lived that story. Their countless letters, diaries, pictures, official documents and records serve as the collective memory of the nation. The National Archives of Canada preserves Canada's archival heritage, and makes it available to Canadians through a wide variety of means – publications, exhibitions, special events, as well as reference and researcher services. Founded in 1872, the Archives' collections today include millions of records, including texts, photographs, films, maps, videos, books, paintings, prints and government files, that bring the past to life. We encourage all members of the public to visit the National Archives of Canada – to pursue a special field of research, to view the constantly changing exhibitions, to learn more about Canada's rich and diverse heritage.
>
> (National Archives of Canada, http://www.archives.ca/)

FORMAL SOURCES

Formal sources of documentation written for public consumption are one of the most useful sources of evidence. If a researcher wants to study the work of a particular business or institution then there is a vast range of official material at their disposal, much of which is written expressly to inform. For example, Felix Driver (1988b) has used the reports and journals of the National Association for the Promotion of Social Science to show how new notions of space and society emerged in nineteenth-century Britain. Such reports were written in the nineteenth century often to inform people about the medical conditions of the urban poor. Driver, however, demonstrates the power of these documents to underpin particular projects of moral reform and environmental improvement.

Perhaps the most impressive formal source is the Census, published every 10 years in the UK from 1801. Other statistical surveys such as the National Food Survey also generate important cultural information. This material is considered separately, however, in Chapter 10. In addition another important source is the reports of various Royal Commissions which have been set up since the 1830s to take evidence and make recommendations on a range of social, cultural and environmental problems. Chris Philo's (1998) research into the exclusionary geographies of tin miners on Dartmoor has used quotations from UK Parliamentary Papers such as the *Report of the Commissioners Appointed to Inquire into the Condition of All Mines in Great Britain* (1864). He uses this report to point out that behind the cultural representation of the rural idyll there is often a harsher reality of poor living conditions. Another potential governmental source is the publications of parliamentary debates, for example those in the Houses of Commons and Lords from 1812 onwards – often referred to as *Hansard* after the man who founded the publication. Parliamentary debates are a particularly useful source for researching previous cultural conflicts. For example, debates over the cultural and economic values of fox hunting in England, can be traced through the centuries using *Hansard*. Debates since 1988 can be found on the Internet at http://www.parliament. the-stationery-office.co.uk/pa/cm/cmhansrd.htm.

Perhaps the most important primary, official and public but, usually, non-governmental documentary source is the press. Daily newspapers can date back nearly 300 years and they give the most important cultural and political views of the time. They also report the general day-to-day events of life. As such, for the cultural geographer they are particularly useful for getting a hold on past senses of place. However, one needs to be aware of potential personal and political bias in newspapers from, in particular, their editors and owners. Nevertheless, local newspapers, in particular, can give valuable information about the way of life in a particular locality if used critically. Newspapers obviously concentrate on what is regarded as newsworthy; the selections and omissions are themselves an indication of contemporary notions of importance and relevance which can give useful clues to the values of places in former times. The recent archives of many newspapers can now be accessed and searched through the Internet (see, for example, http://www.the times.co.uk/); however, older issues are held usually on microfiche in national or regional libraries (see, http://www.bl.uk).

Other potential formal sources are revenue records, court records, church records, trade union records and press/broadcasting records. McFarlane (1970) used a combination of court and church records to demonstrate that witchcraft accusations in Tudor and Stuart Essex were a response to new urban pressures on rural community and kinship structures. Again these have varying conditions of access, but gradually all sorts of formal records are being placed on the Internet thus greatly increasing their accessibility. Some records, such as letters and diaries, however, may not be held in the public domain but may be in private archives. Most of these would generally be thought of as informal sources and are rarely found on the Internet. Access often has to be carefully negotiated with the owner of the material.

INFORMAL SOURCES

I want to begin by examining the informal sources that are written for public consumption. The first of these are the memoirs (accounts written generally for publication after the writer's death), chronicles, biographies and autobiographies that are often produced for posterity. Indeed, more than ever before politicians, military leaders and business executives are intent upon recording their personal and public experiences for posterity. There are also accounts published by less overtly powerful people, for example those in the criminal underworld, and sporting and entertainment celebrities. Similar personal accounts can also often be found in news-papers, magazines, radio and television documentaries, even chat shows. These are often the most accessible of all documentary sources and are valuable mainly because they suggest lines of inquiry. They show how people from particular social groups organize their days and their usage of language and imagery. These particular documents all have their own unique characteristics. Authors are generally interested in showing them-selves in a favourable light. They may have old scores to settle or particular axes to grind. They are written with the benefit of hindsight and are thus subject to the bias of long-term recall. They also tend to focus on the powerful, the extraordinary, the famous and the articulate and are thus particularly useful in researching elite groups in any society.

Memoirs tend to differ from biographies in terms of the degree of introspection given (biographies are more introspective) and in terms of the period covered (memoirs cover a much shorter period). For example, in my own research into the masculinity of British foresters in India I used various hunting memoirs such as Stebbing's *Jungle By-Ways in India: Leaves from the Note-book of a Sportsman and a Naturalist* (1911). Stebbing's book is also a compilation of observations and incidents 'extracted from notes kept in the diaries which are the outcome of sixteen pleasant and interesting years spent in the Indian Forest Service' (1911 p. vii). Although A.J.P. Taylor once described memoirs as 'a form of oral history set down to mislead historians' (Barnes 1988 p. 33) a careful reading of Stebbing's book and other memoirs such as Sanderson's *Thirteen Years among the Wild Beasts of India* (1878) helped me, in particular, to piece together a more convincing description of the everyday opinions and activities of forest officers at work and at leisure than would be given by a reliance solely on other archival sources, such as reports, that only gave me an insight into their world of work. Forest officers, however, as part of the elite civil service administration of India had a tendency to embellish upon the blandness of everyday life. The usefulness of their memoirs thus usually depends upon whether the author has been frank or discreet and to what extent he or she may have exaggerated his or her own role in events. Overall, these memoirs helped me to identify the major concerns as seen through the eyes of leading participants and even gave some indication as to the individual intentions of specific forest officers which could then be assessed against the performance shown in the more official texts (Hannam 2000).

LETTERS AND DIARIES

We also have access to many informal unpublished sources such as letters and diaries. Private correspondence can give an insight into family and social relationships like no other source. In her research, Anna Secor (1999) cites passages from the eighteenth century edited letters of Lady Mary Wortley Montagu, written on her travels in Turkey with her husband the ambassador, to reveal a way of representing Turkey from particular gendered, Orientalist and class perspectives. Diaries, on the other hand, often give a more detailed insight into the lives of people, who can be categorized by their social backgrounds, and tend to reveal more about personality and opinion. They often reflect highly subjective and unconscious responses to events and they often disclose painful self-analysis and frank confessions which would not generally be found in published autobiography.

We should also note that many informal documents are held by ordinary people in their homes as personal archives: the account books of small businesses and charities, the minute books and files of local clubs and societies, timetables, pamphlets and 'flyers', everyday personal correspondence, even e-mails. Indeed, the quintessential cultural activity of shopping produces a relatively large number of documents which may end up in a personal archive of sorts: tickets, lists, receipts and bank statements, all of which when examined reveal something of our preferences, our lives and relationships. There is also a wide range of more visual documentary material such as photographs, cartoons and postcards that have yet to be deposited in public archives but are an invaluable source for the contemporary student. They will often need to be contextualized through interviews with their owners before they can have any real use. The relative importance of informal material depends upon whether or not it survives and whether it accurately reflects a particular way of life or opens up the aspirations of groups and individuals.

TASK

If you have access to the Internet, access one national and one regional archive from the lists below and note for each:

- how far the records go back
- the geographic limits of the archives
- how the documents are held and organized
- the regulations for their access and use
- interesting examples of elite and popular cultural activities
- whether records influence or are influenced by concepts of national heritage.

National Archives

http://www.bl.uk/
http://www.archives.ca/
http://memory.loc.gov/ammem/ammemhome.html
http://www.nmnh.si.edu/naa/index.htm
http://www.naa.gov.au/index.htm

Regional archives

http://www.durham.gov.uk/recordoffice/
http://www.records.nsw.gov.au/
http://www.tsl.state.tx.us/
http://www.state.nj.us/state/darm/archives.html
http://www.bcarchives.gov.bc.ca/index.htm

CONCLUSION

This chapter has examined the range of formal and informal material to be found in archives. We should note the fragility of the material in both public and private archives. Over time, some items get lost, stolen or deliberately destroyed; others merely fall apart, disintegrate or, as I once discovered, get eaten by large rats. It should always be remembered, though, that this material has generally already been selected and sorted, first by the author and second by the archivist. Card-file systems of searching in archives are increasingly being replaced by powerful computer search programmes, particularly as more and more material is placed on the Internet (see Stein 1999). The ways in which a researcher makes sense of the material in terms of selection, sorting and analysing will be considered in Chapter 17.

FURTHER READING

Barnes, J. (1988) 'Books and journals', in A. Seldon (ed.), *Contemporary History: Practice and Method*. Oxford: Blackwell.

Bennett, T. (1995) *The Birth of the Museum*. London: Routledge.

Driver, F. (1988) 'The historicity of human geography', *Progress in Human Geography*, 12: 497–506.

Ogborn, M. (1992) 'Teaching qualitative historical geography', *Journal of Geography in Higher Education*, 16(2): 145–50.

Stein, S. (1999) *Learning, Teaching and Researching on the Internet: a Practical Guide for Social Scientists*. Harlow: Longman.

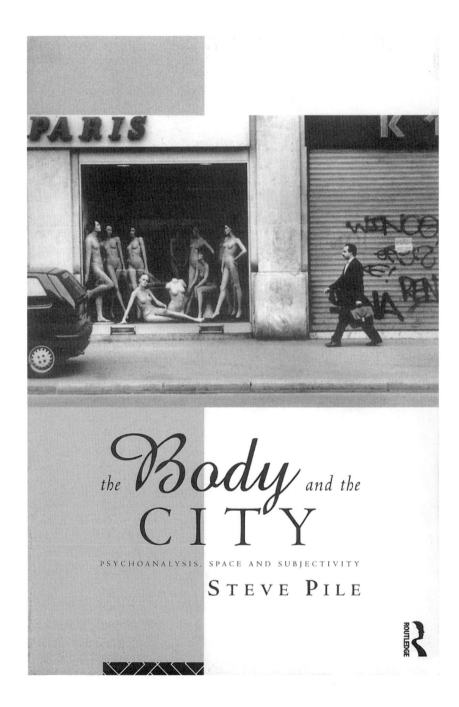

the *Body* and the
CITY

PSYCHOANALYSIS, SPACE AND SUBJECTIVITY

STEVE PILE

12 Reading texts

Pamela Shurmer-Smith

An important part of any cultural system is the construction and consumption of texts. It has become conventional, in the manner of Ricoeur (1978), to extend the term 'text' beyond print on paper to apply to anything with a degree of permanence that communicates meaning. One can, therefore, 'read' a landscape, an urban design, a building, the appearance of a room, a film, a painting, a TV advertisement, a news broadcast, documentary or a soap opera just as well as one can read a book, a leaflet or a map. One can similarly 'read' a spectacular event, an exhibition, a performance or ritual. Implicit in the idea of reading is the assumption that one is making sense out of something experienced: for example, Jenny Robinson's (2000) reading of the 'Cities on the Move' exhibition at the Hayward Gallery is a starting point for a critique of the Western orientation of urban theory.

The extension of the idea of text from mere writing makes additional sense when one knows that the word 'text' comes from the Latin *textere* – meaning 'to weave', as in 'textiles'. A text then becomes a fabric, something which is crafted (Bass 1978). A significant part of this crafting and interpreting lies in the assumption that ideas can be communicated by reference to other, often more concrete, things: a metaphor *suggests* meaning by reference to something else.

Are there messengers? Yes,
space is a body tattooed with signs, the air
an invisible web of calls and answers.
Animals and things make languages,
through us the universe talks with itself.
We are a fragment –
accomplished in our unaccomplishment –
of its discourse.

(from 'A draft of shadows', Octavio Paz 1979 p. 147)

Making sense is an activity. It is not just a matter of absorbing a meaning that was put there by an author; it is, rather, staking out a meaning one can understand for oneself. We can assume (but can never be sure) that this meaning might be convergent with the intentions of the author; but we also know, because we hear authors debate with their critics, that readers can derive more, fewer, or even radically different meanings than the author was conscious of. This does not imply that intelligent authors will be displeased by the discovery of extra meanings received in dialogue with readers. Hélène Cixous, for example, famously enjoys being surprised by radically different meanings from those she thought she was incorporating in her writing (Cixous and Calle-Gruber 1997 pp. 101–2), whereas Umberto Eco (1992) ruefully accepts his loss of control regarding the interpretation of the title of his novel *Foucault's Pendulum*.

The pendulum I am speaking of was invented by Léon Foucault. If it were invented by Franklin, the title would have been *Franklin's Pendulum*. This time I was aware from the very beginning that somebody could have smelled an allusion to Michel Foucault: my characters are obsessed by analogies and Foucault wrote on the paradigm of similarity. As an empirical author I was not happy about such a possible connection. It sounds like a joke and not a clever one, indeed. But the pendulum invented by Léon was the hero of my story and I could not change the title: thus I hoped that my Model Reader would not try to make a superficial connection with Michel. I was to be disappointed; many smart readers did so. The text is there, and maybe they are right: maybe I am responsible for a superficial joke; maybe the joke is not that superficial. I do not know. The whole affair is by now out of my control.

(Eco 1992 p. 82–3)

One is left wondering whether Eco is not deliberately winding up his 'intellectual' readers of his critical work with one more joke. How many rounds of reflexivity is one required to indulge in, and at what point is one supposed to call a halt?

POLYSEMY

An awareness of multiplicity of readings is important to cultural geographers because it permits access to the various elements of understanding that can be generated by people who are able to communicate with each other.

This extract comes from *Livingetc*, a lifestyle magazine. A team was asked to read from a room what sort of person they thought occupied it.

Jamie Oliver
The person who springs to mind when I look at this room is Alan Partridge. This bloke is worryingly organised. He's got lots of files, and although he's clearly quite intelligent, he might be quite anal. He always gets up on time and I bet he works as an Assistant Manager at the Midland Bank. I don't think he's sporty, but he'd probably like to be, which is why he'd have a go at the dumb-bells in the privacy of his own room. He just needs to lighten up – let his hair grow, fray his jeans and get hold of some Calvins instead of Y-fronts.

Jenny Eclair
We all know this type, respectable on the surface, but when you pull up the floorboards it's knee-deep in murdered hitch-hikers. He's only got that iron so he can strangle you with the flex, and why has he got so many white T-shirts? Because he has to throw them away when they get blood-stained, that's why. Plastic suit bags? Body bags more like. Get this killer locked up, that's what I say. You have to be careful with neat blokes. Never go into a man's room unless there's a poster of Pamela Anderson on the wall – those are the normal ones. When you can see the Hoovered carpet, that's when you have to worry.

Phil Hilton
If I was a woman coming back to his place, I'd feel confident that he was clean and his sheets were crisp. The only worry would be that during foreplay he may stop to fold up his clothes as he removes them, which could kill the moment. I'm very shocked by the shoe storage. In my experience shoes are kicked off and then left randomly on the floor – storing them neatly is borderline psychotic as far as I'm concerned.

TASK

- What do you think this room looked like to provoke these reactions? (Imagine the style of furniture, colours, ornamentation.)
- See whether other people visualize it the same way as you do.
- Do you think that the comments of the panel are unique to a particular category of British people?

Few people working in any aspect of cultural geography today would be confident in asserting that they could offer a single authoritative explanation of what anything meant. This should be seen as an indication of the subtlety of the field of study rather than the futility of attempting to practise within it. Whether one is trying to make sense of performances, utterances or material things, one's aim is to reveal as many layers and as wide a spectrum of meanings as one can access. Cultural geographers have to be willing to acknowledge that in any situation there are many truths about places and spaces, some of which are even contradictory. It is largely a matter of personal faith whether one feels one needs to go further than this to believe that all these small divergent truths make up one great multifaceted Truth which constitutes the core of understanding.

> Question, Ganapati. Is it permissible to modify truth with a personal pronoun? Questions Two and Three. At what point in the recollection of truth does wisdom cease to transcend knowledge? How much may one select, interpret and arrange the facts of the living past before truth is jeopardised by inaccuracy?
>
> (Tharoor 1989 p. 177)

TASK

> In *The Great Indian Novel* (Tharoor 1989), the character Ganapati, who is acting as a secretary taking down the narrator's story, does not answer. What do you think?

Steve Pile's (1996) book *The Body and the City* has on the front cover a copy of the photograph by Roger Mayne entitled 'Man and shop window, Rue de Réaumur, Paris'. Pile's preface is given over to his multiple interpretations of the picture, some in the light of friends' reactions to his selection and an analysis of his realization that he is scrutinizing the picture voyeuristically: 'I stepped closer to the photograph, but I still could not work out whether the bodies were people or mannequins . . . either way, it dawned on me that I could be scrutinising a voyeuristic or titillating photograph of naked women (and one man). I felt embarrassed and ashamed.' Pile reads the man in the street as a businessman, 'unaware either of the photographer or of the bodies in the shop window. He seems free in this space' (1996 p. vii).

Pile's reading is gendered. My own reading is completely different. I think that there is a *mixture* of humans and dummies in the window but, more significantly, I am convinced that there is not one man in the picture but two, and that the one in the window is also flesh and blood (the figure on the left). My reading then goes: beauty, youth and playful self-assurance on one side of the picture; undistinguished, conventional, worried middle-age on the other. That the beautiful poseurs are white and tall, whilst the scurrying man in the street is short and probably of Middle Eastern origin, puts another slant on the picture for me. I surmise that the people in the window are making an intellectualist statement, whereas the briefcase contains fabric samples for a small workshop. I think the people in the window have the power. Perhaps I also use and read cities differently from Pile? It would not be helpful to ask which of us had the 'correct' interpretation, for the interpretations come not from the 'reality', which neither of us can know, but from ourselves.

Whatever sources of information one draws upon, one needs to be conscious that people and things always communicate more than they seem to on the surface (hence Katy Bennett's recommendation in Chapter 14 that one attends to more than just the transcribed words of an interview or focus group). This is particularly important when trying to interpret texts as sources of local and temporal meanings of space, place and environment. Although it may be tempting to refer to a text as if it were a fixed version of local knowledge, citing it for purposes of exemplification, texts need to be interpreted in the light of as many local (and relevant external) truths as possible.

INTERSUBJECTIVITY

As we have already seen in Chapter 3, it is not possible to take for granted that everyone understands the world in the same way. However, to save ourselves from going mad with doubt about just how much knowledge we share with others, the concept of intersubjectivity implies the construction of a 'we' who view the world from the same point. This is, of course, an ideal rather than a reality, though it is an ideal which seems at times to come close to realization.

Common experience of texts helps to construct this sense of shared meaning, since texts provide a repertoire of images, narratives, emotions, characters from which people can construct a dialogue. As Geraghty says of prime time TV soaps, shared texts have the ability to 'engage their audiences in the narrative and ... to open up for public discussion emotional and domestic issues which are normally held to be private' (1991 p. 5.) Canonical texts serve as a source of shared wisdom. A tiny phrase such as 'The fault, dear Brutus ...' can conjure up a wealth of meaning about self, responsibility, determination, free will, fate, loyalty for many people who know (or know about) Shakespeare's *Julius Caesar*, but would mean nothing to other English speakers who draw upon a different textual repertoire. Using shared references, people can feel (though not always with justification) that they are drawing close in understanding.

DISCOURSE

We may speak of particular *discourses* marking out boundaries amongst people who communicate with a high degree of consensus not only about the meanings of words and images, but also on a shared agenda. Foucault (1972; 1980; 1981) has popularized the idea that it is through discursive practices that an equation can be made between knowledge and power, that the discourses of the powerful establish their values and opinions as dominant. Discursive practices can be highly exclusionary when only 'insiders' know the special meanings of words, quotations and even shared jokes.

TASK

- Think of phrases from literature, film or song which, because they are widely recognized, summon up a wider meaning than could be derived from the words themselves.
- Decide whether there is likely to be any national or regional variation in the understanding of these.
- Decide whether you think they would mean the same for all categories of people in your society.

We are dealing here with the power of symbolic meaning, but symbols gain particular meaning when they are decoded. The code, however, comes from an ever shifting repertoire of shared knowledge. Because different elements are brought to each interpretation, meanings do not remain constant over time or space. 'Cultures do not hold still for their portraits' (Clifford 1986 p. 10) and discourses are not impregnable fortresses. An important task for a cultural geographer is to show how discursive practices relating to space and place reinforce or resist structures of power. It is in the analysis of texts that particular discourses can be identified and their power revealed.

TASK

- Examine a text from a genre you are not familiar with (this could be a commercial report; a political manifesto from a party you do not support; a teenage magazine etc.).
- Make a note of the words, phrases, references which you could not imagine using yourself.
- Try to identify the tone of the communication and what you understand its underlying message to be. (For example, you may decide that the relentless emphasis on 'fun' in magazines for teenage females makes 'fun' a more serious matter than seriousness. You may then ask why.)

INTERTEXTUALITY

Texts are, themselves, constructed not just out of unmediated experience but also in the light of other texts. Obviously, this is an important part of the formation of shared discourses. One learns to recognize a language of communication whereby, for example, films draw upon devices employed in earlier films, pick up conventions from novels and folk lore, feed into advertising, music videos and fashion. Usually this is a matter of telling new stories in recognizable forms, but sometimes this is done self-consciously, as either a tribute or a joke, where the audience is expected to enjoy recognizing the reference. (The 1992 film *Wayne's World*, directed by Penelope Spheeris, took the playful use of intertextuality and pastiche to such extremes that much of the humour would be lost if one could not recognize the derivation of quite small elements.) However, leaning on old textual references can also result in stale and derivative communication.

> When you park in the city and walk down the hill towards the great red blob in the distance, you want to feel like a diplomat brokering the peace between two warring factions, but you feel like a scab, crossing the picket line. Sunlight blurs the silver pinnacle into further abstraction; it could be a lightning conductor, a crow's nest, a crucifix.
>
> Without doubt, the outside of the building is partly responsible for the mythology that's grown up around it. It's been described as Kafka's Castle, the Alamo and the Kremlin, or as some type of fortress with its narrow windows and ramparts. It also looks a little bit like Guildford Cathedral, the way it looks in the film *The Omen*. Given its position on top of a mound at the end of Leeds's main drag there's also a touch of Capitol Hill about it – seat of power and influence, Big Brother looking down from his perch. Some people have objected to Quarry House because it doesn't fit in with the Leedscape, but quite frankly, that has to be a compliment. The skyline of the city looks like a drawing board, where ideas of all types have been tried and tested, then built properly somewhere else. Closer to, there are statues and sculptures hanging around outside the main entrance, but it's not easy to work out the symbolism.
>
> (Armitage 1998 p. 175)

Simon Armitage is not only an important young poet, he is also a geography graduate. His work is acutely sensitive to the symbolism of space and place, but he constantly approaches this ironically.

TASK

- List the instances when Armitage communicates his impressions of the building by reference to other texts.
- How much sense do you think his description would make to someone who had not received a Western education?

GENRE

The term *genre* is used when there is an expectation of a high degree of conformity to textual conventions. One can talk of Gothic novels, romantic poetry, horror movies, westerns as genres: 'genre' just means 'type'. Here the interest is to operate in a new way with a fairly limited stock of symbols. ('Everyone' knows what it means when a traveller arrives at a castle in a forest, if it is a horror movie. 'Everyone' knows what waves crashing on the shore mean in a romance.)

Although there has been considerable study of film genre emanating from cultural studies, whereby spaces and genders are entwined (Doane 1987; Gledhill 1987; Krutnik 1991), there has been little in the field of cultural geography, where, attention has generally been paid to the places within specific films. There is still plenty of scope for geographers to make their contribution to the understanding of the internal spaces of various film genres. The film critic's crucial concept of *mise-en-scène* (setting) is inherently spatial.

In modern societies people are accustomed to decoding a vast range of texts, many of which reiterate and reinforce each other, on a daily basis. Even when one is not deliberately seeking out textual representations, it is difficult to escape being confronted by mediated images, some of which are barely noticed, others of which we puzzle over. Advertising is, of course, a prime example, as is background radio or TV in public places. We cannot help but learn their language.

REPRESENTING PLACE AND SPACE

Texts depict places and they use space as an element of communication. It is through texts that we imagine places we have never been to, but we also use them to reinterpret those we know first hand. Like many other disciplines, geography has problems with the question of representation – wondering whether it is possible to present a place (or idea) without representing it. The term has two major, related, meanings. One is that of re-presenting, i.e. presenting again in a different form; the other is standing in the place of, i.e. speaking for, the way a lawyer represents a client. When reading a text it is important to be aware of not only what the writer intends in terms of representation, but also what the writer inadvertently communicates. Texts can (and should) be read 'against the grain' (Eagleton 1986), sometimes to reveal the author's view as rather less benign than he/she would wish. Often what is put forward as a simple piece of realist reporting needs to be thought of as a value loaded representation of the ideological position of the writer because it reveals deep assumptions in its choice of words and phrases. There is much useful work to be done on who represents which places in which ways.

Representation is never just cold reporting; it influences the ways in which people encounter spaces and places. Places acquire personalities, reputations, stories as a consequence of the ways in which they are depicted and then react to knowledge of their image. Jacqueline Burgess (1981) was probably the first geographer to take an interest in the way a place acquires a reputation and has demonstrated the way in which Hull has laboured under an inappropriate image. Gargi Bhattacharyya (2000) shows how Birmingham has responded to its stigmatized image in Britain by leapfrogging into an international market, where it has not been trivialized by thousands of cheap representations.

Conversely, there can be quite blatant exploitation of texts when local marketing departments declare 'Catherine Cookson Country' in County Durham and 'Brontë Country' on the Yorkshire Moors, and claim Hampshire for Jane Austen. Shops and tea-rooms put up signs drawing attention to the fact that this 'real' place is the site of another, better known, fictitious place. At a more complex level, there can be a reflexive process whereby the place and its image fuse. One might think of the construction of Paris as a near universal site of romance through films, novels and, particularly, song.

It is interesting to think of the ways in which places become demonized, as in the example given of the (actually fairly ordinary) south Essex borough of Dagenham, which is almost universally depicted as uncouth and antisocial, but is also the butt of jokes about the imaginary 'Dagenham Cocoa Rooms' and the surreal, though serious, Dagenham Girl Pipers (Dogginham, right next to Barking, some distance from Catford). What is it: its ugly Saxon name, or just that it occupies a liminal space between 'real' London and the countryside without managing to be a proper middle-class suburb? Whatever it is, people know to sneer at Dagenham without ever experiencing it first hand.

Overground on the District Line: Becontree, Upney, Barking. You crack jokes and laugh so loudly everyone looks at you. You don't think much of those you call the 'shaggies' going to the Embankment for their Saturday demo against the latest Criminal Justice Bill with their badges and placards and slogans.

East Ham, Upton Park, Plaistow, West Ham – fat boys singing football songs, why's that slag staring? – the wasteland around Bow Creek to Bromley-by-Bow and Bow Road, underground to Mile End and across the platform for the red Central Line. Kiker likes that moment going under the ground when you leave behind all that reminds you of the dreary East and there's no scenery until you walk out into the West. The Central Line had to be red, more exciting than the dull worthy green of the District. Everything fits your expectations. Dagenham Heathway is nothing more than a pretend railway station; Tottenham Court Road with its tunnels and passages and

> multicoloured mosaics is new and bright and the gateway to the other
> world.
>
> (Cornell 1998 p. 29)

Once places or generic spaces acquire generally recognized images, these will be repeatedly reinforced by creators of texts using them as settings which instantly communicate a preconfigured load of meaning. The places and spaces thus used become part of the language of communication.

TASK

> - Watch a 10 minute clip of a video of a popular recent film with a small group of fellow students.
> - Take note of the place(s) in which the action is set, the environment and the use of space. Decide what part the setting plays in the action it supports.
> - What does the setting mean?
> - Could the action be transposed into a different setting without a change of meaning?
> - Compare your reading of the setting of the film with other members of your group.
> - Decide whether a romantic film like *Sliding Doors* could have been set in Dagenham (also on the London Underground) without a radical change of genre.

There is plenty of work for cultural geographers to do looking at the way in which texts create places, attract or repel residents and tourists, influence property prices. All of the authors of this book are, or have been, residents of Portsmouth, a place most English people erroneously think is in Devon and can visualize only through a classic radio comedy called *The Navy Lark*. Others 'know' it to be tough and dangerous through reading the thrillers of Graham Hurley, the blurb for whose most recent book describes the place as 'One of Britain's most turbulent cities' (Hurley 2000). This mercifully preserves us from the hordes of tourists who would doubtless descend if they knew how lovely it is and what a short journey from London. But the trick when analysing texts is to refrain from focusing on revealing the gap between 'reality' and image: there is bound to be a gap or there would be no point in having an image.

SELLING PLACES, PLACES SELLING

Advertisements are some of the most self-conscious texts produced by modern societies. Teams of researchers are employed to try to weigh up every meaning they could possibly convey. The realization that places have images which are textually constituted means that tourist boards and local authority marketing departments work hard to communicate messages which render their places attractive. Jacqueline Burgess (1982) drew attention to the rather bald textual strategies employed in office relocation advertisements in the past to claim that almost anywhere was 'central' (to something), modern, traditional, leafy. But place promotion has become much more sophisticated in the past two decades, aiming to draw attention not just by intelligent advertising but also by ensuring that their places are constantly referred to in art and lifestyle texts (Miles 1997). Cities which successfully attract major spectacular events such as the Olympic Games are conscious that they will be exposed to a global reading and work hard to ensure that it is flattering. As the new millennium was chimed in, only one city in each time zone had a chance to flash its message to the world's TV screens: Sydney and Paris were wonderful but, inevitably, the Thames failed to catch fire.

By the same token, 'readable' places and generic spaces with easily recognized images become foils for commodities. When places are used to advertise goods, the texts that are generated simultaneously reinforce the meaning of the place. One might think of the plethora of car advertisements set in the Scottish Highlands and Islands where both car and place are depicted as rugged, dependable but devastatingly beautiful, or Paris as a setting for perfume advertisements where the product and the city are made to appear feminine, sophisticated, romantic. Advertisements, however, are notoriously easy to subvert and, once one has seen a wittily defaced ad, its original will never again have the same persuasive power. (It is a quite amusing game to take place based advertisements and see how few words or symbols are needed to change their message.)

GENERATING A GEOGRAPHY

Brosseau (1994) accuses the majority of cultural geographers of raiding literature rather than engaging with it. He divides practitioners into two categories – those who look for a literal representation of the outside world and those who look for the 'reflection of the soul contemplating or experiencing this same reality' – but wonders 'why one would resort to novels when more "reliable" sources are available, or when geographers can do their own fieldwork' (1994 pp. 337–8). In his contemplation of the geography of Dos Passos' *Manhattan Transfer*, he attempts to sidestep this by giving literature its own unmediated place within geography: 'We experience another way of writing and producing meaning, of interpreting social diversity and contingency in space, which theories and academic discursivity cannot always accommodate.' Here he is arguing that novels themselves have 'a way of generating a geography' (generating, not depict-ing) (1995 pp. 106–7).

Whatever approach one takes to text – realistic documentation of place, evocation of atmosphere, clues to personality, experimental creation of new emplacements – one needs to remember that every text is generated in the light of grounded experience, subjective view and exposure to other texts. Through this one creates a language of symbolic communication which is rich and ever changing. As postcolonial theorists have amply demonstrated, even apparently apolitical texts such as Jane Austen's romances (Said 1993) or scientific descriptions of scenery (Pratt 1992) can be interrogated for repressive world views. Derrida (1978) has alerted us to the need to read for what is *not* present in a text; this does not literally mean *everything* that is not present, but what might have a claim to inclusion. The act of selection always contains the act of exclusion, and all of the work of creating any text, whether it be overtly fictitious or apparently factual, consists of making selections of style, words, images. It is this textual selection which gives us some of the most powerful clues to any cultural environment.

The abandonment of, and indifference to, the wretched of the earth frac-tures the credibility of [much contemporary] intellectual enterprise. Gibberish – not so much in the terminology as in the substance – attracts graduate students and mature scholars alike, precisely because it is incomprehensible to many and thus exclusive. Articles in scholarly journals read amazingly identically, citation by citation, term by term, paragraph by paragraph. Their attention is almost always rivetted on each other's publications, seldom venturing to the outside world. Ideas rapidly circulate the world over, skimming the scholarly consciousness, which is ever ready for the next move like TNC investments.

(Miyoshi 1997 p. 62)

Miyoshi's view is something the authors of this book are painfully con-scious of, but we also know that if one doesn't speak the language, one doesn't get a hearing. It is a very sobering experience for cultural geo-graphers to interrogate the academic texts they read to discover how little of the world chips into the dialogue of citations, how few texts that were not generated in the West are read by academics anywhere. I have written (Shurmer-Smith 1998) of the problems of intellectual isolation which resulted from immersing myself in academic literature from India as the best way of understanding the contemporary configurations of India. No one much in British academia wanted to talk about works they had not read, in preference to those which were on 'everyone's' hot list. There is a very real cultural geography to cultural geography, and it is worth studying through its texts (particularly the bibliographies).

RESISTANCE AND REACTION

It might seem that deconstructing texts is a rather insipid thing to be doing, when one contrasts it with the personal immersion necessitated by practical fieldwork. But texts can be tools of change. Poetry and song have long been weapons of resistance and revolution: Che Guevara had with him a copy of Pablo Neruda's *Canto General* when he died, and Neruda himself spent many years in exile from his native Chile because his poetry was regarded as dangerous (Neruda 1991; 1978). Anna Akhmatova destroyed her poetry but had it committed to memory by her friends, and its power lasted beyond the Stalinist purges (Akhmatova 1976; Cixous 1994). Both use haunting descriptions of their homelands to evoke their very different ideologies.

All over the world, people who are striving to transform their societies put considerable effort into the production of texts, because it is through communication that they can forge alliances and gain support. The Zapatista movement has famously constituted itself as much through its use of the Internet as it has through local struggle. Organizations such as Lokayan in India exist to provide an exchange of information between small local movements and diasporic groups, and are renowned for their almost obsessive text production. Both left- and right-wing movements internationally are defined by their publications, websites and videos. Charities and religious groups purvey their ideologies through texts, not always revealing themselves in the light they would wish to. The similarities and differences in styles of communication of differently motivated movements internationally shed interesting light on ideas about global culture, but I have yet to see it researched.

Textual analysis is worthwhile, almost regardless of the nature of the texts selected. It can be an activity of a few minutes or can be the basis of a full-scale piece of research, but either way, it should always strive to render as many meanings as possible, and not be taken as a detective's clue to a single truth.

GLOSSARY

Discourse The communicative conventions of a group or category of people. So 'academic discourse' is not usually acceptable around the family dinner table.

Genre A type (of text) (from French *genre*). Genres can cut across media, e.g. a thriller can be a film, a radio play, a novel.

Intersubjectivity Having (apparently) the same subject position (seeing things the same way).

Intertextuality Recognition that texts draw upon each other.

Medium Means through which communication is made, e.g. print, radio, film, Internet (plural 'media').

Narrative The way in which a story (including a factual account) is told.

Polysemy/polysemous With a spectrum of meanings.

Reading An interpretation.

Text Anything fabricated by human beings that can be 'read'.

FURTHER READING

Barnes, T.J. and Duncan, J.S. (eds) (1992) *Writing Worlds: Discourse, Text and Metaphor in the Representation of Landscape*. London: Routledge.
Duncan, J. and Gregory, D. (eds) (1999) *Writes of Passage: Reading Travel Writing*. London: Routledge.
Hawthorn, J. (1992) *A Concise Glossary of Contemporary Literary Theory*. London: Edward Arnold.

13 Participant observation

Katy Bennett

Participant observation attempts to understand the everyday lives of other people from their perspective. It requires researchers to situate themselves in the lives of others and to allow their voices and actions to influence the research agenda. To study cultures, subcultures, and the value systems and social structures which make up these, researchers work within the patterns of relationships in a given setting. This method embraces a range of fieldwork experiences with researchers adopting different roles and levels of participation depending upon the demands of the research context. This chapter does not offer a smooth, clean approach to participant observation, but attempts to show its possibilities and opportunities to cultural geographers. Experiences of participant observation come in all shapes and sizes and no story of fieldwork is the same. It is important to realize this from the start. Everyone has different ways of approaching, coping, seeing, listening, feeling and writing. Participant observation can be an exciting research method when it brings the different voices of people to inform theory and uses their knowledge to transform practice.

Participant observation is a method that attempts to get beyond forcing answers to questions framed by a particular knowledge. It is about trying to open oneself up to different ways of constructing social life and knowledge. This implies not only learning to unravel and interpret behaviour and coded meanings, but also striving to think and behave in a way which makes sense to the people one is living with. Judith Okely (1983) explains the strength of participant observation and shows why structured interviews were not useful to her research on gypsies. She found that they could quickly sum up the needs of a questioner and respond in the most appropriate way to get rid of them fast with their 'ignorance intact' (1983 p. 45).

Invariably the questioner assumes an authoritarian role, and in any case little opportunity is given for volunteered information and insights unpredictable to the observer. Although Gypsies may be an extreme case for the inappropriateness of questions, caution should be exercised in fieldwork anywhere.

> Beliefs and explanations for actions may only be explicit among the actors in certain contexts, and not as answers to questions. Some of the explanations given may be mystifications. Ultimately the total meaning may never be articulated by the group participants, but instead be the work of the anthropologist to unravel.
>
> (Okely 1983 p. 44)

Participant observation owes its heritage to anthropology and the method is layered with images of men and women dedicating long periods to fieldwork in alien cultures and inhospitable conditions. Participant observation, though, is used closer to home by people outside anthropology. It is a method that has been used by people working in cultural geography, but influenced by anthropology, exploring various issues such as the cultures and spaces of clubbing (Malbon 1999), the Indian Administrative Service (Shurmer-Smith 1998), people with mental health problems (Parr 1998b), the workplace (Crang 1994) and recreational living history (Crang 2000). Sometimes people are explicitly covert during parts of their research and do not inform those they are observing of their study (Cook 1997; Parr 1998b), while others attempt to be more overt, but this is not always clear cut as researchers inevitably only partly reveal their intentions. Researchers also organize their participant observation work in different ways, spending extended periods in a place (Shurmer-Smith 1998) or aligning participant observation with their lives when their research is time and place specific (Crang 1994; Malbon 1999). Inevitably, though, participant observation does not end when 'the field' is left and 'lives' with the researcher as s/he stays in touch with her informants, sometimes returning, and is affected by her experiences.

Participant observation can be a bit of a contradiction because it is about both taking part and observing. Too much emphasis on observation might undermine a fuller understanding of the researched, whereas staunch participation might result in 'what is called "going native", that is . . . over-identifying with and being an uncritical celebrant of the subculture' (Thornton 1997). Getting the balance 'right' is sometimes difficult and demands the researcher navigate the study around difficult issues of ethics. Fuller (1999) found himself increasingly involved in the running of a credit union, which he set out only to observe. He accounts for how political motivations drove his research into a different sort of study whereby he played a much bigger role in a situation, which he was also simultaneously attempting to study. Without his involvement, the credit union, which he came to care deeply about, might have failed. Fuller (1999) managed his research in a way that satisfied both his academic and his political needs. There is no necessary reason why participation should be restrained, committed or withdrawn; it is all participation and needs to be monitored by the researcher. One needs to learn to adapt and balance participant observation in unique ways which suit personal and academic demands.

SHAPING 'THE FIELD'

Although participant observation is about attempting to understand other people's worlds in their everyday setting, there is nothing natural about this particular method. We construct a space where we can do our research and talk to those who can help us. Katz writes that 'to have these conversations in a way that is distinct from everyday life, we must have a "field" marked off in space and time' (1994 p. 67).

Theoretical concerns and political motivations underpin participant observation and affect the construction of the field, shaping the aims and hunches of the researcher. Hunches are not about proving or disproving hypotheses or tunnel visioned objectives, but are theoretically motivated issues with which the researcher plans to engage whilst doing her field-work. Hunches are not smooth and neat, but open to surprises, change and even transformation through the progression of the fieldwork as the researched people (hopefully) play a role in the direction of the study.

'The field' is a construction as its setting is deliberately chosen and delineated; the researcher's aims, ideas and motivations are layered onto it and s/he affects 'the field' with her presence. The people s/he is studying alter their behaviour accordingly, as does s/he. The field is a space of betweenness (Katz 1994) because 'it is not the unmediated world of the "others", but the world between ourselves and the others' (Hastrup 1992 p. 117). Better still, it is an 'interworld' (Crossley 1996) where the subjectivities of the researched and the researcher become entangled as they bring their worlds to bear upon it.

Because of the influence that the researcher has on her fieldwork, it is important that s/he is reflexive and aware of herself (and her behaviour) and the ways in which s/he influences her study. Our relationships with the people we research, how we write through the research and what we write push us to confront issues of ethics and to question both ourselves and the research. Such questions 'are not so much arguments against ethnographic fieldwork as a case for ensuring that we remain reflexive and critical' (Coffey 1999: 74).

Reflexivity is often misunderstood as 'a confession to salacious indiscretions', 'mere navel gazing' and even 'narcissistic and egoistic', the implication being that the researcher let the veil of objectivist neutrality slip. Rather, reflexivity is self-critical sympathetic introspection and the self-conscious analytical scrutiny of the self as researcher . . . A more reflexive and flexible approach to fieldwork allows the researcher to be more open to any challenges to their theoretical position that fieldwork almost inevitably raises.

(England 1994 p. 89)

Supporting a reflexive approach to participant observation is an awareness of positionality, a self-knowledge of potential performances both inside and outside the field.

We are differently positioned subjects with different biographies, we are not dematerialized, disembodied entities. This subjectivity does influence our research.

(England 1994 p. 92)

In very simple terms, positionality is the standpoint of the researcher which is affected by her biography (history, experiences, age, sex and so on) and influences her study. It should not be treated unproblematically as a thing to be staked out at some point in the research, but should be taken as dynamic and relational depending on the research context.

TASK

- Read and make notes on the following extract from Jenny Diski's *Skating to Antarctica* (1997).
- Why do you think that the travellers will reproduce different versions of the place?
- If you are doing this in a class, get into small groups to compare your impressions.

The most offensive person on the trip was a Scandinavian professional photographer, getting shots for a new book. He elbowed people out of his way, commandeered swathes of space, stuck his camera into the creatures' faces for the big close-ups, and ran busily everywhere in search of the telling shot. He was as single focussed as a camera lens, but some of the unprofessional others came close. It wasn't just still photography, of course. The camcorder was much in evidence, so, added to the click and whirr of motorized snapping, was the monotonous murmuring of voices, not people in conversation with each other, but individuals talking into their machines, adding commentary to their motion pictures. Every time I heard what I thought was someone talking behind me and politely turned to listen, I saw a Cyclops with video camera replacing the missing eye, pacing deliberately about, moving the machine and their head up and down and around, as if eyes no longer swivelled in their sockets, muttering into their chests. To anyone not aware of the purpose of the camcorder, we would have been mistaken for an outing of the deranged. Of course, people were not actually talking to themselves, or even thinking aloud, but talking to their friends and family at some time in the future when they would be sitting in their living rooms watching the video. The present experience was already in the past for them, they had skipped over time, and were seeing the world through their video lenses, as it would look when the current moment was dead and gone. Things were named and described, sentences formed, a final draft written, without the first-draft struggle to transform wordless impressions into language. There was no translation of world into words, just the direct commentary, cutting out all the processes that might have added up to reflection. The memories created now would exist, frozen in the future as lens-framed news reports.

(Diski 1997 p. 153–4)

PLAYING 'THE FIELD'

Part of the reason why there is nothing natural about the field is that we play a part in its shaping and there is nothing natural about our behaviour. In the words of Judith Okely, 'Any suggestion that the anthropologist merely be "natural" or "herself/himself" is unrealistic' (1983 p. 42).

Whether or not they are doing participant observation, people behave differently in different settings according to where they are, who they are with and their agendas. People's behaviour when they are with their parents is usually different from when they are with their friends. Like actors, most people are capable of a number of different performances depending on the occasion. When doing participant observation, researchers continually work on their behaviour and performances so that they might blend into their research contexts. Participant observation can be physically challenging, with researchers using their bodies in unfamiliar ways and subjecting them to new experiences.

It was important that I walked without purpose and authority, also that I sat slumped, often staring into space for hours at a time, as this was the norm at the drop-in. To produce money to buy the endless cups of tea that people drank in the centre, it was necessary to have small change (never notes) in my pocket. To have a wallet, purse or larger amounts of money would have been different, organized in a way that the majority of people were not.

(Parr 1998b pp. 31–2)

To assist their performances, people might dress in a particular way to help them feel included and part of the scene. This is similar to researchers and how they prepare for the field. Researchers give considerable thought to how they look, smell, sound, so that 'others' might relax and be less suspicious of them. Such considerations help to shape the performance.

TASK

- Read the extracts in the next two boxes and compare the different ways in which two researchers present themselves whilst doing their fieldwork.
- Imagine a research situation you might generate and then consider what management of self would be needed.

When accessing a drop-in for people with mental health problems in a deprived inner-city location, it was appropriate for me to attend in dirty, old clothes. What seemed important was not only what I was wearing, however, but how I presented my physical body (while walking, sitting, standing, drinking, eating and talking), and also more ambiguous considerations, such as how I smelt. At first, I would attend the drop-in in appropriate dress, and then be uncomfortably aware of how my hair smelt of shampoo. The people with whom I was sitting might not have washed their hair for months, and shampoo set me apart from the other people in the centre. In later visits, I learned not to wash before I arrived, perhaps not as a means of ensuring intimacy, but certainly in order to limit some of my own 'otherness' in that situation. The ambiguity of the body (how we adorn ourselves with soap, scent and deodorant to disguise real bodily odour) was highlighted in this research setting, and my body, in this sense, had to become much more physically present.

(Parr 1998b pp. 31–2)

During my own fieldwork I was extremely conscious of the need to manage and produce an acceptable body to the fieldsite. I was concerned with

> presenting a personal front which mirrored that of the social actors in the field. I attempted to dress 'like an accountant'. This in itself was based on my assessment of the acceptable body in accountancy – as smart, self assured, confident and well managed. I dressed in a black interview business suit with straight skirt and fitted jacket rather than denim jeans or Lycra leggings. I wore pale blouses rather than T-shirts in loud colours. I wore heeled shoes rather than training shoes or loafers.
>
> (Coffey 1999 p. 65)

FIELDING POWER

Participant observation always involves imbalances in power. Sometimes these seesaw when researching elites and those who powerfully perform particular parts of their identity, but normally the researcher is in a position of power as it is s/he who defines 'the field' and is prepared for its end when the research is done. The researched have been at the centre of a study for which they sometimes get little in return, whilst the researcher walks away with not only diaries and notebooks that contain information, but also, at times, the hopes and aspirations of the researched.

> Fieldwork is inherently confrontational in that it is the purposeful disruption of other people's lives. Indeed, anthropologists even speak of the 'violence' of fieldwork, even if the violence is symbolic. In fact, exploitation and possibly betrayal are endemic to fieldwork.
>
> (England 1994 p. 85)

WRITING 'THE FIELD'

An essential part of participant observation is writing: writing through/out the research as often as possible. Whilst Chapter 19 deals with creating texts, I want to think briefly about writing whilst still in 'the field'. Working on the task suggested below might help to explain the problems.

TASK

Consuming food is a cultural as well as a biological activity. Participating in and observing people's consumption practices can reveal social structures and value systems. This task introduces you to writing field notes and research diaries. These are written in different ways depending upon the needs of the researcher and the constraints of 'the field'.

> For a week, using the method of participant observation, keep a diary on all aspects of food and eating in your household, hall of residence or college. Take note of:
>
> • spatial and temporal organization of eating, such as who sits where, how the room is arranged when food is eaten, the reasons for these
> • power relations regarding food, its preparation, serving and eating, observing who controls and dominates these
> • how a household/institution identifies itself through its food.
>
> What would a stranger have to learn to fit in at times when food is prepared and eaten?

Phil Crang (1994) worked as a waiter in a restaurant to research workplace culture. He found that the need to work efficiently as a waiter made recording his observations difficult: after all, when people go out for a meal they do not normally expect to become the subject matter of someone's PhD.

> Field notes were taken on my order pad when possible (this was not possible when very busy, so then I wrote single 'scratch notes' and elaborated them in the break period at the end of the shift) and these were written from to produce shift-by-shift research diary entries (usually written through the morning after a shift, given that an evening rarely finished before 1.30 a.m.). The latter included initial notes of 'factual detail' (that is, an expansion of field notes) followed by deliberately speculative reflections on these.
>
> (Crang 1994 p. 676)

TASK

> • Compare your own experiences of writing a research diary with Crang's.
> • Explain the reasons for any differences.

Diaries usually contain three elements. These are *field observations*; *attempts to make sense of these*; and *feelings in relation to the research*. Not only does the presence of researchers affect their research, but so too does how they feel when they participate, observe and write their diaries: this influences interpretations and understandings. Field notes and diaries show the unfolding of a study as it develops, with the researcher negotiating potential obstacles whilst also opening herself up to new issues.

When writing a diary, there are inevitably moments when there is concern as to what is being written, what should be written and how all of this writing will be of any use. This was probably experienced when doing the task and writing a diary that is (hopefully) rich in detail regarding food matters. The problem with the task is that it has no clear theoretical motivation, resulting in a loose framework within which to work. For example, if feminist theories motivated the research, there could be concern for issues relating to gender relations and food and the sort of household that these create. Similarly, a poststructuralist agenda could dismantle assumptions based on the nuclear family and mealtimes as represented through, for example, the advertising of food products. To show this, there would probably be a need to get together a group of individuals who have completed the task to discuss findings and produce some sort of presentation where results and conclusions can be revealed.

Not only is the writing process itself fraught with problems and concerns of what to include, but also the circumstances of writing can be problematic and sometimes guilt-ridden. Have you told those in your household, college or hall of residence about your research? Do you write notes in front of them or do you write them elsewhere? Do you let other people read your research notes and research diary? Field notes and diaries are often written away from the gaze of the researched, and there are inevitably moments when the researcher feels guilty for writing about those with whom they interact 'in the field'. Worse still is the realization that research might accentuate the vulnerability of the researched. What issues do you face in your research if you come across someone with an eating disorder?

LEAVING THE FIELD

Participant observation requires considerable involvement in the everyday lives of the people being researched, which sometimes makes leaving 'the field' hard to do. Occasionally, farewells are more difficult for those being left behind because they might not have been taking steps to distance themselves from the researcher in the same way that s/he may have been in relation to them. Sometimes guilt on the part of the researcher sets in. The following extract is from my PhD thesis and captures my curious mix of emotions upon leaving 'the field' when living and working for a farming household for board and lodging only for four months in exchange for their taking part in my research. I was caught between the sadness of leaving, having become involved in their lives, and the practicalities of continuing to observe events and 'use' them for my thesis right up until the moment when I drove away.

I cried when I left the household because, over the four months, I had grown close to its members, particularly Karen and the children ... On the last night, at my leaving supper, I set up a video camera in the corner of the room to record the event, so that I could later analyse it from a different perspective; I was still doing my research whilst they demonstrated their affection. The presents which I received were given to me in recognition of the work that I had done for the household and with them came the feeling that suddenly our 'bargain' had slipped in my favour and I had 'used' them more than they had me. My diary, notes and video recordings were a guilty burden that I drove out of the field with and jostled in boxes alongside my gifts and cards, which carried messages of staying in touch, and that I would be greatly missed.

(Bennett 1998 p. 40)

The researcher though is often deeply affected by participant observation. Performances that are at first unusual can begin to own the researcher, with the research experiences writing themselves indelibly upon selves and bodies.

There are also unconscious ways in which the fieldworker adapts and more fully participates. You learn through the senses and in the body. Posture and movement synchronise with those around. There is a photograph of me standing with a Traveller, both of us posing for the Gorgio cameraman. I have assumed unconsciously the exact stance of my companion: arms folded, making a defensive barrier.

(Okely 1983 p. 45)

CONCLUSION

Participant observation is not the method to use for a short-term research project. The one-week food consumption task was designed simply to introduce you quickly to an important part of participant observation, the writing of field notes and diaries. It uses a situation with which you are already familiar and will have few problems in accessing, where you already know people. For these reasons it works as a short-term exercise, but no more. It skips over some difficult stages of fieldwork such as the delineation of 'the field' and, in particular, accessing people in strange places. The first part of fieldwork can take quite a lot of time (a month is nothing) whilst the best ways of getting to know places and people are thought through: mistakes based on over-enthusiastic blundering in may be impossible to retrieve.

It is important, however, to know what shapes participant observation so that publications based on the method can be interrogated. Participant observation is saturated with all sorts of issues which affect its use. These issues have been introduced in this chapter and emerge as 'the field' is created, relationships develop and the researcher is aware of her behaviour and the effects of her study on the researched. All of this influences the substance of an ethnography.

Participant observation can be a difficult method to use for an undergraduate dissertation where time is a constraint. It is not impossible, however. Ian Cook (1997) provides examples of dissertations using participant observation that he has supervised. Places where you already have connections in terms of work or leisure could be developed into interesting research topics in relation to workplace cultures and subculture. A group of people which listens to and identifies itself through particular music at specific venues and shares knowledge through specialist magazines and websites might be best researched over a period through the method of participant observation. Participant observation work lasting several months can be incorporated into timetables of other activities, yielding plenty of material for dissertations. It is also a method that can be used in conjunction with others discussed in this book, such as interviews and focus group work.

FURTHER READING

Coffey, A. (1999) *The Ethnographic Self: Fieldwork and the Representation of Identity*. London: Sage.

Katz, C. (1994) 'Playing the field: questions of fieldwork in geography', *The Professional Geographer*, 46: 67–72.

Parr, H. (1998) 'Mental health, ethnography and the body', *Area*, 30(1): 28–37.

Lokayan
BULLETIN

● *Unethical Medical Research*

● *Plastic Devastation*

● *How to Combat Corruption*

● *Peoples' Global Action*

March-April '98

14.5

 **14** Interviews and focus groups

Katy Bennett

Whilst some methods expose broad patterns using aggregate data that normalize out, interviews and focus groups are two methods that creep up close to a particular point of the pattern to engage with individuals who make up statistical averages. Done well, all methods have their use, but the aim of focus group work and interviews is to expose differences, contradictions and, in short, the complexity of unique experiences. Both interviews and focus groups take many different forms depending upon the aims, background, skills and theoretical perspective of the researcher. There is, however, a fundamental difference between interviews and focus groups: interviews usually involve the researcher and one 'other', whilst focus groups include the researcher and a group of people willing to take part in the study. There are features, though, which are common to both, with the researcher having to work through difficult issues of power and control. Interviews and focus groups are not easy methods to use and often present researchers with tricky obstacles which they must carefully navigate. Before features common to them both can be explored, consideration needs to be given to what makes interviews and focus groups different from each other.

FOCUS GROUPS

Focus group work is a good method to use when time is a constraint: hence its evolution through market research. It is also useful when working with communities to understand their histories, responses and thoughts in relation to particular issues, such as coalfield communities in relation to pit closures. Through the research of Burgess et al. (1988a; 1988b) focus group work in cultural geography has moved forward to embrace psycho-therapeutic traditions (see also Burgess 1996). Unlike the task set out below, which serves only as an introduction to focus group work, Burgess et al. (1988a; 1988b) advocate focus groups that meet more than once. Their groups met five or six times, allowing relationships to develop between individuals that then influenced the content of conversations and played a role in shaping the direction of the research.

> It takes time for people to begin to explore beneath these well-rehearsed phrases and feelings, to acknowledge their doubts and fears about the natural environment, and to move beyond a superficial consensus towards an exploration of the diversity of environmental experiences and values within the group.
>
> (Burgess et al. 1988b p. 457)

Focus groups require the researcher to work with a gathering of individuals who sometimes know each other. Although the researcher has issues that s/he wants the group to discuss, her aim is to get the group as a whole to shape understandings and knowledge with individuals interjecting, agreeing, disagreeing, verifying and so on. The researcher tries to encourage people to speak in their own ways, using 'locally relevant terms' which are familiar to them. During focus group work, the researcher also attempts to explore group dynamics as individuals dominate and take control of the situation and others allow this to happen (or not). What may take shape in focus groups is 'the formation of a temporary social structure that is a microcosm of the larger context' (Goss 1996 p. 118).

Focus group work is a method increasingly used in a variety of contexts by those working in cultural geography. Focus groups have been used by Robyn Longhurst (1996) to explore women's experiences of pregnancy; Beverley Holbrook and Peter Jackson to research the 'many meanings that people invest in their everyday shopping practices' (1996a p. 138); Claire Dwyer (Laurie et al. 2000) to study young British Muslim women's identities; and Jackie Burgess, in some of her more recent work, to study 'social and cultural understandings of fear in recreational woodland' (1996 p. 130).

There are inevitable obstacles to face when using the method of focus groups, and some of these will have to be faced in the task I am setting. Deciding who to include and ways in which to reach them can have repercussions for the research findings. Researchers have used groups of people who know each other as well as people who are strangers to one another. They have also considered the make-up of groups in relation to gender, age, class, ethnicity and so on of individuals involved. Researchers find different ways of reaching potential participants: using local activists, contacting those in positions of responsibility, and attempting themselves to recruit individuals through local newspapers and posters displayed in community centres (see Holbrook and Jackson 1996a). The process of recruitment can be problematic, something that is reinforced by Lily Kong's (1998) research experiences in Singapore. She was working not only against a research context that 'subscribes to positivist positions' (1998 p. 82), but also with people 'for whom public participation is not part of the culture' (1998 p. 81). The reluctance of individuals to get involved for a variety of cultural and personal reasons can lead to 'failed' focus groups (Longhurst 1996). Robyn Longhurst ended up with only two people in a couple of her focus groups, although she discovered advantages in this in relation to her study on pregnant women. Kong (1998) argues that the difficulty of getting a focus group to come together at all means that sustaining multiple meetings can be a hard task.

Early on in research, consideration needs to be given to the management of focus group meetings. In recent focus group work with teenage girls, I and another researcher found that we had to work really hard to get discussions going at the start of the session and to encourage quieter members of the group to contribute. In retrospect, we cut in too quickly when there was a silence. Silence may mean that the respondents have said what they want for the time being, or are thinking of something else to say, or have stopped talking because of what someone else has said, or are quite used to lapsing into silences when together; this needs to be given consideration. Knowing how to work with silences is as important as knowing how to manage talking, and takes confidence and skill.

Focus group work needs to be organized so that responses and discussions might rock the suppositions of the researcher, but also work within the broad themes and aims of the research without getting hijacked by other people's agendas or turning into counselling sessions. Organizational issues are not helped by a lack of information on how researchers run their focus groups (Kong 1998), and this is a shame because it has repercussions for research outcomes. We know how they ideally should be organized, but experiences of the method are hard to come by. Returning to the work of Burgess et al. (1988b), some insight into how they organized and managed one focus group is provided. They show the topics that they encouraged the group to cover in each session and how they managed discussions so that individuals focused on the relevant issues and did not end up providing a support group for each other in relation to other, personal matters. They also draw attention to the importance of preparing group members for the end of the research, when the special and particular dynamics of the research situation must come to a close.

TASK

- Organize six to eight people to meet to discuss the effectiveness of *The Big Issue* in combating social exclusion and the implications of 'an alternative to a dependency culture'.
- Give them a copy of the extract below and ask them to read it before the group meets.
- Prepare themes that you want the group to discuss and starter questions to initiate discussions.
- Take the role of leader of the focus group, which you should allow to run for no more than an hour.

- Tape record the focus group.
- Write notes on the meeting as soon as it has finished. (Make sure that you include details of how individuals interacted with one another and the influences of these on the discussion.)

At this point you ought to transcribe your tape, including an indication of pauses and stumbling, but this is a very time-consuming activity and is not worth doing in full in the context of an exercise. Transcribe the first five minutes to get the experience.

- Identify the main problems you experienced.
- Decide whether you think that this is an effective way of gathering data.

The Big Issue, a paper sold by the homeless, is not a new idea. Since the 1980s, *Street News* has been sold on the streets of New York by the homeless and the jobless. Gordon Roddick, chairman of the Body Shop, brought the idea back to London and it eventually saw the light of day in the summer of 1991.

It was very simple then in its initial approach. It never set out to change the social fabric of Britain. It was modest. Sold on the streets, the vendor would get the lion's share of the cover price. There were no tricks. Money went directly into the homeless people's pockets rather than being thrown to a third party to dole out later.

Our first problem in *The Big Issue* was that unless the public wanted to buy the paper, then it wasn't a legitimate deal. If they were buying it as a pity purchase, then it was just a hidden hand-out. It couldn't simply be about homelessness because you could not build a large enough readership, and no large readership meant no income for the homeless. *The Big Issue* had to be something that homeless people wanted to sell and the public wanted to buy. We expanded the arts pages, put in personality interviews and tried to make it work as a good read as well as a social read.

We gave homeless people an opportunity to earn their own living and to stand on their own two feet; it was about empowering people and giving them an alternative to dependency culture.

(Bird 1993 p. 10)

INTERVIEWS

Interviews usually involve the researcher and an interviewee, although this is not fixed. Sometimes other people such as family members and colleagues disrupt the interview and the research situation is a sort of hybrid of interview, focus group and participant observation. Like focus groups, interviews are organized according to the needs of the researcher but should be flexible, allowing informants to play a role in the shaping of the research. Gill Valentine writes that 'Interviews, in contrast to question-naires, are generally unstructured or semi-structured. In other words they take a conversational, fluid form, each interview varying according to the interests, experiences and views of the interviewees. They are a dialogue rather than an interrogation' (1997a p. 111). This means that, whilst you should have a checklist of the issues you would like to cover, you should be prepared to let the encounter run its course, as this will not only allow the interviewee to express herself in her own way but also raise matters you might not have anticipated. This is not always easy to manage. To prevent the conversation from drifting too far from what you are interested in, you need to keep pulling it back with phrases like 'I was really interested in what you were saying about . . .'. Most people find that these encounters work best if the interviewer also reveals something of herself (but it is important not to dominate).

Interviews are the chosen method when researchers are keen to hear the 'stories' of individuals in a situation where they are (probably) unin-terrupted and uninfluenced by the presence of other people. Researchers use interviews when studying sensitive topics and vulnerable people (for example Valentine's 1993 work on lesbian experiences and perceptions of everyday spaces, and Parr's 1998a work on people with mental health problems). Interviews benefit research that aims to capture individual experiences: for example, McDowell (1997a) used them to study employees' experiences of workplace interactions in investment banks in the City of London, and Massey used them to explore 'how particular dualisms may both support and problematize certain forms of social organisation around British high-technology industry' (1998 p. 157). Interviews make a researcher sensitive to differences and contradictions, adding to the rich-ness of data. The task set out below aims to familiarize you with the process of interviewing and encourages you to compare it with focus group work.

TASK

- Ask one person to read the same passage as you used for your focus group.
- Set up an interview with that person.
- Confront similar issues to those you raised in the focus group.
- Do not let the interview last more than an hour (it could well be much less).
- Tape record the interview.
- After the interview, write notes on your experience of the encounter.

You ought now to transcribe your tape, but skip that stage because of the time needed.

- Do your findings from the interview differ in nature from those obtained from your focus group?
- If so, in what ways?

Interviews are often held in places which are familiar to interviewees, such as their home or workplace or public places known to them. Such venues often give further insight into the lives of the researched, but might also affect what they are willing to discuss or how they respond to questions. Consideration to the context of the interview is important.

CONTROLLING POWER

Interviews and focus groups are loaded with complex power relationships. Hopefully you began to get a sense of these whilst doing the tasks suggested. Power relationships are not fixed or unidirectional, but shift and change according to how the researcher and researched are interacting with one another. Structural inequalities and personal differences based on class, status, age, race and sexual orientation all affect the research situation, and are played out in different ways both consciously and less consciously. Whilst many researchers are sensitive to power imbalances because of their control over the organization and dissemination of the interview and focus group, there are times when informants play up parts of their identity to make the researcher feel uncomfortable. This happened to Cotterill (1992) when interviewing mothers-in-law who occupied apparently higher class and status positions than she and used these to put her in her place.

Most researchers are, however, only too aware of their control of the research situation and the way in which this powerfully affects interviews, focus group work and their research in general (Miles and Crush 1993; Katz 1994; England 1994). Oakley (1981) is often quoted for her pioneering attempts to challenge the power politics of the research situation when interviewing women experiencing motherhood for the first time. Aware of textbook procedures for interviewing that encourage distancing from and the objectification of the researched, Oakley attempted to subvert the usual one-way flow of information from interviewee to interviewer. When questioned about her own experiences of motherhood by her informants, Oakley gave of herself so that friendships developed. Her approach to interviewing has created considerable debate and helped to reveal knottier problems of methodologies around the issue of power and control.

Cotterill (1992) argues for friendly relations rather than friendship when discussing her research on affinal relationships between women. Whilst interviews and focus groups can produce moments of closeness, with informants opening up to the interviewer and the interviewer seeming to do the same, the researcher is able to walk away from the research situation with her tapes and notes. The researched might not be able to walk away so easily. Ultimately, researchers need to be aware of the ways in which they play on their identity, using different facets of their experiences to engage with their informants in the best possible way. Oakley (1981) played up her own experiences of motherhood to her informants but was less up-front about other aspects of her identity.

Despite the attempts of researchers to reduce their control of interviews and focus groups, writings and recordings mean that their powerful roles need to be recognized during the research process. In the end, it is the researcher who attempts to capture the words of the researched so that s/he can make sense of them later. Whilst researchers and researched inevitably both understand and misunderstand each other in interview and focus group situations and have to work hard to produce meaningful encounters, it is the researcher who will analyse what was said, how they interacted. It is the researcher's (mis)understandings that will be represented.

TAPE RECORDING

Managing tape recorders can be an awkward task. The machine sits between researcher and researched and its presence undoubtedly influences interviews and focus groups. Its effect depends on the individuals involved. Sometimes it can help to define the research situation, protecting it from disturbances and other aspects of people's busy lives and providing space for relationships to develop and experiences to be discussed. When I taped interviews with farmers' wives in their homes, if children or husbands walked into the room with requests and questions they would often point at the machine and tell them to go away because they were being interviewed. In other situations the tape recorder can be intrusive, muting the researched and restricting what they feel able to recount. Tape recorders can also inhibit the researcher, with her gaze flitting between the researched and the machine as s/he (re)checks that it is switched on and working. S/he can also feel self-conscious about recording her questions and comments if s/he is part of a research team and others will scrutinize what is recorded. (Maybe some of this relates to your own experiences of using a tape recorder when doing the tasks?) Transcribing always takes much longer than is suggested by methods books because of background noise, quiet voices and people talking over one another.

My idea of using unstructured taped interviews as the main method of data collection changed. In Namibia, the popular theatre activists I did manage to conduct taped interviews with felt uncomfortable and intimidated by the process. Although it is true that these interviews were conducted too early in the research process to really allow for a relaxed atmosphere, I began to wonder about the reliability and exploitative nature of the taped interview process, even with people more familiar to me, in situations where the power balance was so unequal. This specific point was articulated further as one popular theatre activist consistently postponed any interview times I would set up. After two months, he finally declared that he did not want to be interviewed on tape. He found the process exploitative, felt he did not want to devote the time to it, and did not understand why I wanted to conduct a taped interview given that I already knew all the information he would describe through our informal discussions.

This resistance to being taped is understandable given who I was and my direct approach which progressed too quickly. Others experienced it as intrusive. Consequently, the taped interview process was actually exploitative and demeaning; one popular theatre activist told me that I was researching them 'like they were animals in a zoo'.

(Farrow 1995 p. 78)

'I asked to use a tape recorder, but she looked really freaked out, "I'd rather you didn't really. I get really nervous. When I was in psychotherapy, the woman made me talk into a tape recorder, I mean forced me. I dropped it and everything and I couldn't talk, but she made me sit there holding it".' (extract from research diary, 27/4/94)

The taping of interviews is notoriously difficult and obviously affects the telling of a story and the narration of the self. A key consideration was that many psychiatrists and other mental health workers use tape recorders in their discussions with clients, which means that tape recorders become associated with a patient–professional relationship which is power-laden and imbued with issues of access to care, medication and possible psychiatric abuse. Taping an interview was also problematic in my study as some interviewees often experienced states of mind which incorporated paranoid thoughts about being monitored, recorded and bugged.

(Parr 1998a pp. 347)

Taking written notes can be equally difficult because they also do nothing to subvert power relations. Whilst some people are unaffected by note-taking, others slow down or stop talking whilst you write to 'help' you or because you have suddenly made them feel self-conscious. Writing scratch notes during the interview and then writing them up immediately after (or as soon as possible) is sometimes a good alternative. No matter which way you choose to capture the words of the researched, it is always a good idea to write up the focus groups or interviews in detail as soon as they are over.

INTERSUBJECTIVITY

When interviewing someone or running a focus group, nervousness might mean that energies are devoted to keeping the flow of conversation going and less to noting how you are interacting with each other. This might seem odd to read, but nerves can get in the way of thoroughly listening and observing, which can affect responses to the researched and the direction of the conversation. Few people admit to this (or maybe it is only I who have to work hard at these skills) but Sue Wilkinson (1998) shows how researchers do not make the most of focus group work.

> For this article I reviewed well over 200 focus group studies ranging in date of publication from 1946 to 1996 ... Focus group data is most commonly presented as if it were one-to-one interview data, with interactions between group participants rarely reported, let alone analysed. Where interactions between participants are quoted, they are typically used simply to illustrate the advantages of focus groups over other methods, and analysed solely at the level of content (rather than in terms of their interactional features).
>
> (Wilkinson 1998 p. 112)

Whilst listening to informants you observe them, and their behaviour influences your performances during the interaction. They also observe you and each other and do the same. This is what creates intersubjectivity, a sensuously relational experience (Crossley 1996) whereby people consciously and unconsciously construct their own meaning, objectify 'others', recognize themselves in them and play on their performances accordingly to engage as best they can. Intersubjectivity is dependent upon reflexivity and positionality and affects the research situation and ultimately its findings.

ISSUES OF ETHICS

Intersubjective encounters inevitably have ethical repercussions.

> It seems that the acceptance of subjectivity, involvement and interpersonal relationships in the research process is likely to raise difficult ethical questions for the researchers . . . Feminist-inspired notions of doing research 'with' or 'for' rather than 'about' women (or other 'others') seem admirable and are becoming widely accepted within human geography (at least to those who hold to a notion of emancipatory geographies). However, it is becoming clear that the adoption of qualitative or ethnographic methods alone does not release the scholar from exploitative relations, or even the betrayal of her subjects.
>
> (McDowell 1992 pp. 406–7)

> I have emerged from interviews with the feeling that my interviewees need to know how to protect themselves from people like me.
>
> (Finch 1984 p. 80)

By exposing the impossibility of the cool, detached researcher and the inevitability of entangled subjectivities and developing (un)friendly relations, the researcher must confront some complex ethical concerns which have to be worked through. The research situation that the researcher plays a role in creating both consciously and unconsciously provides a context for particular revelations. Sometimes the person who is made into the object of the research feels able to divulge personal details to the researcher, raising problems as to how to handle the data in a way that suits the researcher's needs but which does not have harmful repercussions for the researched. Perhaps s/he promises anonymity, s/he promises not to publish parts of the research, or s/he promises nothing but never feels able to publish the depths of her findings. Researchers have to find their own ways of coping.

So far I have focused on situations when the research has gelled and flowed, and both the researched and the researcher have apparently liked each other and created friendly relations. There are situations when the researcher does not like what she hears and dislikes the researched (Reay 1996). Whilst doing the tasks set, you might have felt these sorts of emotions in relation to other people's opinions of the homeless. How did you handle what you regard as unpleasant opinions and not jeopardize the research situation? Sometimes people deliberately bait the researcher just to provoke an angry response. Often it is only with the benefit of hindsight that understanding is achieved; researchers learn from their experiences (Cotterill 1992; Parr 1998a).

Much of this chapter has been influenced by feminist debates on methodologies which engage with the issues of power and control that saturate focus group work and interviews. Researchers influenced by such approaches aim to expose the marginality and subordination of 'others' caused by the undermining of their experiences, knowledges and ways of knowing. They aim to give informants a certain amount of control over the research context. Hopes of empowering informants sometimes underpin the objectives of such researchers. Whilst Longhurst's focus groups meant that the pregnant women she worked with asked one another questions and 'treated each other as "knowing subjects" ' (1996 p. 147), conversations did not necessarily lead to an empowering research context. Using qualitative methods that attempt to challenge the researcher's control of the research context does not automatically eradicate potentially exploitative relations.

> Women in the small group discussions that I facilitated were coming to full term pregnancy for the first time – this can be an exciting but also a daunting time. I cannot be sure how 'empowered' the women felt at the end of sessions or even if this is a realistic aim. Perhaps, because the method depended upon human relationships – engagement, and even after just one meeting, some attachment – it placed the research subjects at risk of manipulation by myself and by the other woman she talked with in a way that other methods that open up rather than reduce the distance between the researcher and her subjects do not do.
>
> (Longhurst 1996 p. 147)

As observed by England, 'Reflexivity can make us more aware of asymmetrical or exploitative relationships, but it cannot remove them' (1994 p. 86).

Whilst empowering research is a noble aim, this chapter has exposed some of the thornier issues that reflexive and transparent methodologies expose. I am not encouraging a return to the objectivity of positivist approaches, just an awareness that most research benefits the researcher more than it does her informants. Sometimes to even presume to empower implies 'contestable notions of domination' (McDowell 1992 p. 408).

The aim of this chapter has been to introduce the methods of focus groups and interviews and related methodological concerns surrounding the issues of power and control of which researchers need to be aware. Its intent is not to scare but to show that such methods are not an easy option. They need careful planning. In an attempt to explain their differences, I have discussed the methods separately. This does not mean that they should not be used together in a research project. Many researchers have successfully used both, with the methods complementing as one makes up for the limitations of the other. Similarly, using one, the other or both alongside methods introduced in other chapters can also be beneficial.

FURTHER READING

Baxter, J. and Eyles, J. (1997) 'Evaluating qualitative research in social geography: Establishing "rigour" in interview analysis', *Transactions of the Institute of British Geographers*, 22(4): 505–25.

Burgess, J., Limb, M. and Harrison, C. (1988) 'Exploring environmental values through the medium of small groups. Part One: Theory and practice', *Environment and Planning A*, 20: 309–26.

Burgess, J., Limb, M. and Harrison, C. (1988) 'Exploring environmental values through the medium of small groups. Part Two: Illustrations of a group at work', *Environment and Planning A*, 20: 457–76.

Cotterill, P. (1992) 'Interviewing women: issues of friendship, vulnerability and power', *Women's Studies International Forum*, 15 (5/6): 593–606.

Holbrook, B. and Jackson, P. (1996) 'Shopping around: focus group research in North London', *Area*, 28(2): 136–42.

Parr, H. (1998) 'The politics of methodology in "post-medical geography": mental health research and the interview', *Health and Place*, 4(4): 341–53.

15 Field observation: looking at Paris

Louis Shurmer-Smith and
Pamela Shurmer-Smith

One of the things that prospective students almost invariably inquire about a department of geography is where it goes on fieldwork. Fieldwork is regarded as an essential part of the education of a geographer, a time when the books are left behind and students get on with unmediated observation of the real world. It does not always live up to expectations.

Gillian Rose has described undergraduate field trips as, 'The initiation ritual of the discipline' (1993: 69). Then, citing Hart, 'Geographers, like the mythical giant Anteus, derive their strength from contact with the earth . . . even though it may occasionally mean taking risks, living dangerously' (1982 p. 24), she ties the field experience to the testing of masculine virtues and sees it as exclusionary of women's ways of experiencing the world. The majority of human geography fieldwork is, however, nothing like this heroic depiction; it consists of large groups of students being shepherded around urban areas. There are few sadder sights than that of 30 young people (with clipboards) straining to catch the words of a lecture given on the corner of a street in a great European city. There are few things more pointless than the mapping exercises that follow when the students break down into small groups for project work in a country where they cannot speak the language. Fieldwork is an important part of the culture of geography and the form it takes can reveal a great deal about departmental structures of knowledge and distributions of power.

CLAIMING YOUR OWN FIELDWORK

Chapter 13 has already considered ethnographic fieldwork based on participant observation. A short fieldclass is never going to be able to achieve much that is worthwhile in terms of participation, but it can be the ideal setting for developing one's powers of observation and training oneself to ask sensible questions. The first step is to ensure, well before setting off, that the person designated as the leader of the fieldclass has set the event up in such a way that students are empowered. In the field, people should be enabled to see for themselves. The senior members of a fieldclass should be encouraged to regard themselves as part of a dialogue, facilitating the experience by making suggestions and responding to observations, stating their own point of view without privileging it, setting things straight when they recognize factual errors, but, otherwise, trying to restrain the urge to dictate. The greatest value of an experienced member of staff on a fieldclass is that (s)he can indicate short-cuts to the various aspects of the life of the city, but if the field has to be seen entirely through the eyes of a lecturer, one might as well stay back at the university and watch videos (s)he has made. One might ask why there should be fieldclasses at all (people could make individual studies of foreign cities in their own time); we feel that the defence of fieldwork is the shared dynamic, coming together to compare experiences and breaking down into pairs or small groups to discover more. The group experience breaks down the sense of alienation which makes it take time to know a strange city, but the price is that one is burdened by familiar constructions of knowledge.

A fieldclass should not just be a thing in itself, it should act as the fulcrum of other learning. All the teacherly activities should have taken place in the classroom well before setting off, so that students already know the shape of the city, its historical, economic and political background, its major recent initiatives. There should have been plenty of anticipatory reading, not just geographers' studies of the city, but locally produced planning strategies, local place promotion publications, histories, novels, poems, newspaper articles, travel writing and even travel guides. If there are any available, films set in the city should have been watched. Students should be familiar with the basic statistical data on their city before setting off; this in itself will generate useful questions. Most European cities have websites in several languages and one can use these to get well beyond a list of tourist attractions and hotels. Being there will make all these fall into place, and first-hand observation will allow one to flesh out the bare bones of the information. People see so much better if they already have an idea of what they are looking at, and background knowledge will help guard against making naive, ethnocentric, interpretations of what is before one's eyes. For example, if one already knows that a district has a particular composition in terms of class, ethnicity, household structure, one will read the district's physical characteristics more sensitively.

Following a fieldclass there should be plenty of opportunity for reading further and discussing new ideas thrown up by the experience. Fieldwork should open up new areas of inquiry and it is a pity if, instead, it is regarded as the culmination of a programme of learning, with nothing but a short write-up period after returning to the university. Because of the constraints of the academic calendar, fieldwork often comes just before an assessment point, at the end of a term or a semester. Assessment should not mean that the experience is shut down and packed away; fieldwork should be designed so that its insights are capable of being followed through in the next round of courses. Students need sometimes to question the cultural constructions which have evolved in their departments, because, all too often, they have grown up through processes of local expediency and then been maintained by inertia.

This chapter will focus on fieldwork in Paris, a city visited by a very large number of British geography departments. It is also much visited by students of architecture, art, art history, town planning, cultural studies and tourism, and it is worthwhile seeking out other students in one's university to gain other perspectives. This is not intended to be a chapter about Paris; it is, rather, about doing cultural geography in Paris in the context of a short fieldclass.

PARIS

Baudelaire claimed Paris as the capital of the nineteenth century, Benjamin (1979) announced it as the capital of the twentieth; perhaps it has been toppled for the twenty-first, perhaps not. Culture is, of course, everywhere, but Paris has become pre-eminently associated with the idea of culture (with both a large and a small C) in a way that few other places are. Cultural geography in Paris? One might ask what other kind of geography could there possibly be; the place has provided so many changing symbols of urban life that it is virtually impossible for any Western educated person to set aside a host of images and impressions. Paris has certainly not been unaware of this special status (recognizing itself as the 'City of Light') and cultivates its role as the high point of Western culture: the capital of the great kings of France; the site of Europe's first enduring modern revolution; the home of philosophers and artists for centuries; a city of lovers. The more recent presidents of France and mayors of Paris have quite consciously laid claim to a mixed legacy of elitism and populism, history and hypermodernity. The city promotes itself as the pinnacle of sophistication, the last word in fashion, the ultimate in gastronomy, and the epitome of romance. It is very easy to be swept along by the hyperbole. Paris is a wonderful place to work out how so many abstractions can be captured by one city.

We have said before that there is nothing very clever about spotting the existence of a gap between image and reality: images exist because reality is impossible to grasp. There is certainly no point in trying to catch Paris out as a fraud by revealing the ordinariness of the suburbs, the poverty, the racism, the dirt under the disguise: only the most naive tourist would expect it to be otherwise. Indeed, writers have long counterposed the grandeur and the degradation. For us, the excitement in studying Paris, or any other city, is to be able to articulate the complex relationships between the multiple practices and ideologies as they configure to make a place. This may well mean explaining the glories in terms of the deprivations or by reference to the vaingloriousness of politicians. The fabric of any city is a mixture of historical legacy, modern planning, economic and political pressures, popular usages and resistances. The old survives only if the new values and renews it. Every city is in a constant state of flux and it is interesting to decide why the vibrant city of Paris has been so very faithful to the spaces of its past – the modern intellectual industries still on the left bank, the commerce on the right. There are traces and remnants lovingly preserved everywhere but without any concession to living in the past. It is appreciating this flux which makes fieldwork worthwhile; catching at the changes and interpreting them is the ultimate reading exercise.

PUBLIC SPACE

It is difficult to know where to begin with Paris. Should one stand on the Île de la Cité and contemplate the Gallo-Roman settlement of Lutetia and think about the temple of Jupiter underneath Notre-Dame? Should one think how Paris spread out onto the two banks of the Seine, spiritual on one side, secular on the other? Should one remember how it continued spreading to take in the villages beyond the successive walls, those settlements whose activities were not controlled by the city, where the prostitutes and the petty criminals lived? Should one then consider how the suburbs and the new towns evolved? Should one contemplate the kings in their palaces, moving from the Louvre to the Marais and out to Versailles, then the revolutions of 1789, 1832, 1848 and the Paris Commune of 1871 which made the people of the city acutely aware of the power of popular protest, but which also imposed a city plan designed to crush revolt? All of these are important, but we intend to start where any enthusiast for the city starts: walking the streets, then sitting at a pavement café, thinking how, despite its climate, Paris is a place where life is lived in public.

As John Ardagh (2000) points out, much of the affordable accommodation in central Paris has long been cramped and inconvenient; this has forced social life out of homes into public spaces. Parisians of all social classes go out a great deal, the wealthy to the theatre, the opera, the art galleries, the concerts and the fashionable restaurants, the poor to the cheaper bars and restaurants and free public spaces. Public space works in Paris because people have a tradition of using it. Small green havens and sitting-out places have been carved out in the newest phase of city beautification, where car parks are being moved underground and the spaces released are planted with trees and flowers. New parks have recently been created out of old industrial land at Bercy, La Villette and the site of the old Citroën factory. There is plenty for a cultural geographer to explore in terms of the ways in which public space is used in Paris: the spaces of children's play, the unlikely spaces appropriated by skateboarders, the spaces of picnicking and promenading, the open-air markets, including the huge flea market at Clignancourt, the spaces where old people sit and watch, where differently composed ethnic groups meet to chat, the places where beggars sleep.

It is hard not to notice that, with the exception of shopping and the family use of parks, much of this public open space is dominated by men's activities (and inactivities). Paris is still a *flâneur*'s city, but Doreen Massey (1991) points out in criticism of David Harvey that there does not seem to be a feminine form of the *flâneur* – that women cannot idle, gaze about them and appropriate public space in the same way as men. A gendered cultural geography of the public spaces of Paris makes an interesting study. A good starting point for this is Catherine Cullen's (1993) *Women's Travel Guide to Paris* (which, incidentally, refers in its introduction to the *flâneuse*).

The open space has all been planned; some, like the Jardin du Luxembourg, goes back to the pleasure gardens of the aristocracy, but plenty is new. The Parc de la Villette is not just the reclaimed site of the meat market and abattoir relocated to the north-east of Paris, it is also one of the many *grands projets* whereby late twentieth-century Paris has made itself into a site for adventurous architecture and urban landscaping. The projects have all been the subject of competitions, attracting the world's best architects. The entries are widely publicized internationally in serious newspapers as well as the architectural press and competitors display not only their designs but also their philosophies of space. The Parc de la Villette, designed by Bernard Tschumi, is best understood with the philosophy of Deleuze and Guattari (1988) in mind. Unfenced, it melts into its surroundings; it is difficult to decide whether the Canal Saint-Denis is in the park or whether it acts as a boundary or as a conduit. The space is laid out on a strict grid, punctuated by bright red pavilions, but it looks as if there is randomness. It contains concert halls for rock and classical music, the Museum of Science and Industry and the spectacular Géode used for multimedia presentations situated in one of the best free play spaces imaginable. It is a place for activity rather than for the idler and is hard to imagine a park more different from the gravel paths and formal designs of the Jardin des Tuileries or the Jardin du Luxembourg.

HEROIC ARCHITECTURE

At Bercy the City of Paris has attempted to balance the natural westward drift and eastern decline by relocating the Ministry of Finance to a huge new building literally on the Seine, which its architects, Chematov and Huidobro, claim to be a metaphorical bridge and a gate. Next to it is the bunker-like Palais Omnisports (a spectacular indoor performance venue) with its sloping walls covered in grass, merging into the lawns of a new park, occupying the site of former wine warehouses. The park, like La Villette, is unfenced and democratic, but its many paved pathways down to the river preserve the lines of the runways for the barrels of wine unloaded here when this was an industrial site. Across the river is the Bibliothèque Nationale de France, which, it seems, one is supposed to dislike, perhaps because many people think that national book collections should be housed in quiet and retiring buildings. These are all public spaces, public utilities, public investments and grand gestures in a formerly run-down area. Remember to wonder where the people who used to work here have gone.

Paris is a feast of architecture using poststructuralist concepts of space. The Institut du Monde Arabe (designed by Jean Nouvel) announces itself as a dialogue between Arab and French culture and is a favourite of ours. With its perforated metal floors, reflective surfaces and terrifyingly fast open lift, the building has been known to intimidate grown men by it's wilful use of vertical space. Its windows are like the shutters of sophisticated cameras and react to light, but manage also to evoke the latticework of Islamic architecture without any sense of pastiche. From the high terrace the views are stunning.

The Opéra Bastille (designed by Carlos Ott) was intended to entice in people who were intimidated by the classic Opéra. At the site of the Bastille, the symbol of French revolutionary spirit, it is supposed to be a people's opera house and was designed not to overwhelm. Its entrance hall houses a Metro station and shops, in an attempt to democratize what is often seen as an elite activity. Inevitably, much the same people as formerly went to the opera continue to be the main patrons.

The formerly controversial, but now much loved, Pyramide du Louvre by Pei is the visible part of the Louvre underground extensions which added 25,000 square metres of space to the museum. The project was Mitterrand's pride and joy, but many thought, until it was completed, that this ultra-modern glass construction in the Louvre courtyard was an act of vandalism. In fact it acts as a mirror which not only reflects but illuminates this palace of the *ancien régime*.

Few students will miss the *quartier* known as Beaubourg, dominated by the now refurbished Pompidou Centre (designers Piano and Rogers) which once seemed startling for its external utilities, its bright colours and particularly its escalator, from which one gains an ever expanding view across Paris to Montmartre. This is a space of cafés and street performers, of sitting on the paving, a place of petty hawking and dealing. It is a few paces to Forum des Halles, the major Metro and RER interchange, a dreary attempt at an exciting shopping experience on the site of the old wholesale vegetable market. One of us remembers the famous ritual of onion soup and a glass of Sauvignon before dawn when Les Halles was Les Halles, 'the belly of Paris'; the other only imagines she remembers it as a consequence of reading Genet's description in *Our Lady of the Flowers*.

La Défense, in the west at the end of the Voie Triomphale or Grande Axe, is not just von Spreckelsen's Grande Arche (with another wonderfully vertiginous lift) but the unashamedly modernist site of the European headquarters of several multinational companies, the Elf and Fiat buildings being the most spectacular. The CNIT exhibition centre is a great vaulted roof without exterior walls, which challenges conventional ideas of built space. Underground there is a major transport interchange; behind there is social housing with a large North African population.

The *grands projets* may be understood not only in terms of a late modernist statement, but also as symbols of state and commercial power. Though they give one a glimpse of the sublime, they are also awe-inspiring in the same way as the Basilica of Sacré-Coeur, planted on the site of the commune to remind the masses where authority lay (Harvey 1985). These are cathedrals in secular form. If architecture is gendered, this is masculine, in the heroic mode.

All of these new spaces and more can be taken in on foot (if one is fairly hardy) and they need to be set conceptually alongside the older versions of Paris, the Paris of the revolutionaries, the *demi-monde*, the impressionists, bohemians, surrealists, existentialists, situationists ... which people still hope they can visit by trampling the same stones. These spaces certainly reinforce the bourgeois Paris of business, fashion, the arts and shopping.

THE ARSENAL

The Atelier d'urbanisme (an adjunct of the City Planning Office) at the Arsenal mounts excellent temporary and permanent exhibitions relating to most aspects of Paris. It is also the home of an archive of materials on the city (not all in French) to which there is free access, good working space and helpful, multilingual staff. This contains just about everything you might want to consult from planning documents, statements by architects, files of newspaper clippings classified by topic, through survey data to maps and posters, and is an essential early stop for any cultural geographer.

SUBURBS AND NEW TOWNS

But in the enthusiasm for the spectacular spaces of late modernity, one should not forget the suburbs and the new towns. One should ride out to the *banlieues* (peripheral suburbs) on the RER (Réseau Express Régional) to discover where people with families really live. Mathieu Kassovitz' (1991) film *La Haine* depicts the hopelessness of North African young men of the inner suburbs of north Paris and prompted other young directors to turn their attention in the same direction (Konstantarakos 2000).

Paris has tried many times to solve the problem of housing its masses and it has set famous architects from Le Corbusier onwards to the task of designing new utopias. Sometimes it seems as if the *banlieues*, with their high-rise *grandes ensembles*, and the new towns of the Paris region are a series of failed experiments, but France's urban population grew so fast after the Second World War that speedy solutions were needed. Noin and White (1997) are a good source on the Paris agglomeration, particularly the growth of the new towns such as Marne-la-Vallée, Evry, St Quentin en Yvelines. Each of these has attempted distinctiveness of design in its social housing, such as Ricardo Bofill's postmodern pastiches of a classical Roman amphitheatre at Marne-la-Vallée and a sugar-pink lakeside Versailles at St Quentin en Yvelines. There is plenty of scope for deconstruction of these landscapes. There is also plenty of scope for thinking about the lives of the people who commute and those who are left behind.

Your Paris, I thought, was American.
I wanted to humour you.
When you stepped, in a shatter of exclamations,
Out of the Hôtel des Deux Continents
Through frame after frame,
Street after street, of Impressionist paintings,
Under the chestnut shades of Hemingway,
Fitzgerald, Henry Miller, Gertrude Stein,
I kept my Paris from you. My Paris
Was only just not German. The capital
Of the Occupation and old nightmare.

(from 'Your Paris', Ted Hughes 1998)

LIFESTYLE

We have already remarked that Paris is self-conscious about its cultural status, that its promoters play upon the idea of the city both as a custodian of the high arts and as in the front rank of the new and experimental. They also exploit the image of Paris as the ultimate in fashion and lifestyle. A cultural geographer on a short field trip will never get round all the major museums and galleries, let alone discover the small and specialist ones; still, the point of fieldwork is not so much to conduct a piece of original research as to find out well-known things for oneself. Whatever else is omitted, a visit to the great department stores of Haussman's Paris is essential: they are not just a shopping trip, they are a reminder of the dawn of shopping as an experience (Laermans 1993). If you loiter in the tea-room in the cupola at the top of Printemps, you cannot help but be pulled back to the times when imperialism and new industry caused an explosion of spectacular consumption. If you are lucky there will be an in-store fashion parade featuring the distribution lines of the new couturiers, globally recruited but still needing the Paris cachet for their credibility. It may seem as if little has changed, but in the nineteenth century the applause was not for people with names like Issey Miyake.

> Some of the myths of Paris have been deliberately created, others are the product of accretions through time, and have become very well established. Most of these myths have some relation to realities, but others have developed to an extent where they have taken on a life of their own. All, however, influence the ways in which Parisians, and visitors to the city, view the urban space and make use of it. Given this predominance of the myth in the 'reading' of contemporary Paris, it is perhaps not surprising that it has been French, and explicitly Parisian, philosophers who have been at the forefront of the development of semiotics . . . and of deconstruction.
>
> (Noin and White 1997 p. 207)

REPRESENTATIONS

One of the ways of approaching Paris through its cultural imagery is to set oneself the task of tracing out the spaces of writers' representations of the city. We have already mentioned Genet's Paris which was nocturnal and homosexual. Then there is Henry Miller's rampantly heterosexual version of the American in Paris and Anaïs Nin's (1967) arch counterpoint to this. Perhaps there is no more to be said about Walter Benjamin's *flâneur* and the Paris *passages* (Benjamin 1979; Buck-Morss 1989), but they are worth a nostalgic glance as one considers their legacy. One can wonder how bustling Montparnasse could have been the home of existentialism, and then one can visit Sartre's and Simone de Beauvoir's stark and unembellished twin gravestones and remember.

> For poets, Paris is not just Paris, the city of superlative monuments: from Notre-Dame celebrated by Péguy to the Eiffel Tower of Cendrars and the Pont Mirabeau of Apollinaire. It is a city charged with the remembrance of revolutions, of those who struggled for liberty or the Liberation, of the movements of the crowd, of the joys and pains of so many labourers and craftsmen, of immigrants and refugees, each of whom has given their identity to every poet – whether it be in verses of love or nostalgia – they cannot escape the grasp of history and the everyday.
>
> (Dobzynski and Melik undated p. 10, tr. Pamela Shurmer-Smith)

Alex Hughes has written of how, in Simone de Beauvoir's *Memoirs of a Dutiful Daughter* (1963), 'Paris emerges as a "masculine" space onto which the conventional, binary male/city:female/interiority gender divide is indubitably mapped', whereas in Violette Leduc's work Paris is depicted as feminized and maternal (Hughes 1996 p. 118). De Beauvoir's *The Mandarins* (1960) reveals in fictionalized form the Paris of the existentialists, as does Sartre's *Paths of Freedom* trilogy (1947a; 1947b; 1950), and from these another view of Parisian women intellectuals emerges. One might look at contemporary Paris through the lens of the women writers who have helped make its reputation as a place for liberated women: George Sand, Colette, Marguerite Duras (and Simone de Beauvoir herself) would all be good starting points.

Adrian Rifkin (1996) maps out a homosexual literary Paris and reminds us that this city, famous for its egocentricity and uncaring stance, has long been a haven for all people who do not easily fit into mainstream society. In the cemetery of Père-Lachaise one finds the grave of Oscar Wilde, who found refuge from British censure in Paris.

The city has become associated with the avant-garde. The surrealist Aragon (1971) published two vignettes on places in Paris as part of his attempt to write a novel without plot or character, and catches an enduring image of place in his description of the Passage de l'Opéra. He, however, begs the indulgence any student might when trying to capture a place.

> I had taken the precaution of choosing a landscape that would quickly become unverifiable for the simple reason that the passage was about to be demolished in favour of an access way to the Boulevard Haussman. Not so I could tell lies with greater impunity, but rather to conceal my apprentice status as a descriptive writer and so ward off the derision that I feared my efforts might provoke.
>
> (Aragon 1971 p. 14)

The situationists (Debord 1971; Buck-Morss 1989), who trace their roots to both the surrealists and the events of Paris 1968, lie at the heart of the new anarchy which informs so much contemporary global protest. If you share their world view, it is in this city that you can follow up their ludic critique of late modernity. The situationists claimed that they wanted to become what they called 'psychogeographers' of 'the precise laws and specific effects of the geographic environment . . . on the emotions and behaviour of individuals' (Debord, quoted in Plant 1992 p. 58). They were, of course, joking; they wanted to do no such thing when they conducted their *dérives* (drifts) through Paris. They were closer to Benjamin's *flâneur* but with a sense of the ridiculous.

The city is also the home of poststructuralist philosophy, which seems natural in a place which so readily accepts the volatility of late modern culture (Sturrock 1998) and incorporates it into its urban design. The University of Paris in its now fragmented parts has provided a home to Deleuze, Cixous, Derrida, Lacan, Foucault, Kristeva, Baudrillard, Latour – the list could go on. Although it is conventional to lament the demise of the intellectuals in Paris, it remains an intellectual's haven.

The danger with a city whose artistic and literary products have travelled far is that outsiders can be trapped into trying to visit places that are gone. Anaïs Nin's houseboat is no longer moored on the Paris quays; much of the Quartier Latin consists of Greek restaurants; Montmartre has plenty of people painting in the Place du Tertre, but they aren't artists. The places have come to signify different things now, things which people catch at in their desire to remember sentimentalized versions of times they were too late to take part in. Some think that tourist-dominated spaces are inauthentic, but it is interesting to ask whether they are not in fact very authentic examples of new cultural and commercial forms.

Certainly no one could deny that the touristic Paris intercuts artistic, fashionable and commercial Paris. Urry's (1995) ideas about the consumption of places easily can be tested out in this city where everything can become a commodity. The tourist overlay of Paris makes fieldwork less problematic than in many other places. Paris is used to being gazed upon and has become nonchalant about cameras and notebooks. If all you come back with is a boring essay about retail outlets, it will certainly not be the city's fault.

FURTHER READING

Clébert, J.P. (1992) *Les Hautes Lieux de la Litterature à Paris*. Paris: Bordas.

Cole, R. (1999) *A Traveller's History of Paris*. Morton-in-the-Marsh, Gloucestershire: Windrush Press.

Noin, D. and White, P. (1997) *Paris*. Chichester: Wiley.

Sheringham, M. (ed.) (1996) *Parisian Fields*. London: Reaktion.

16 Feminist methodology

Carol Ekinsmyth

> [T]he western industrial scientific approach values the orderly, rational, quantifiable, abstract and theoretical: feminism spat in its eye.
>
> (England 1994 p. 81)

The concerns of feminists, outlined in Chapter 6, ordain a particular stance *vis-à-vis* the research process. The feminist critique of the institution and practice of 'science' has been the basis of a critique of positivism and its associated research methods. In particular, positivism's claim to objectivity, the underlying belief that the researcher can detach *himself* (*sic*) from the research process and act as an impartial observer, has been the basis of a feminist methodology that to the contrary, expresses 'faith in the legitimacy of subjectivity' (Christopherson 1989 p. 87). Related to this, other building blocks of feminist methodology are the acknowledgement of the partiality of knowledge, a sensitivity to power relations, faith in 'everyday knowledges', openness to a diversity of approaches and emancipatory goals for research outcomes. These have been highly influential for cultural geographers whose subject matter and theoretical concerns oblige a similar methodological stance. This chapter will explore each of these themes.

SUBJECTIVITY AND REFLEXIVITY

Underlying positivism and the scientific method (methodologies that have been hugely influential in the discipline of geography and continue to provide the foundations for quantitative methods) is the belief/claim that this detachment is possible. On the basis that it is far more realistic to acknowledge the impossibility of the detached and objective researcher, feminists have developed, and are continuing to develop, a methodology that carefully considers the role of the researcher and the researched (and the relationship between the two) in the research process. In considering the subjectivity of researcher and research 'subjects', feminists have needed to theorize the nature of subjectivity. In this they have been influenced by poststructural thought, particularly the work of Michel Foucault who has claimed that subjectivity is a product of discourse and power relations, not the other way around. Judith Butler (1990) has suggested that performance is the key to conceptualizing subjectivity and identity. Instead of possessing an innate, invariable subjectivity, individuals perform identities/ subjectivities, and these performances are shifting in nature and are regulated by circumstance or context. In this way, space is central to subjectivity as performances occur *somewhere*.

TASK

- Imagine you are undertaking research on elderly people living in an inner city area.
- Do you think that you could really construct a project, interpret the results and write up the findings in such a way that enables you to remove yourself, your influence, your history and positionality from the research process and the research findings?
- Can you really claim that any other person doing the research would achieve exactly the same findings as yourself? (What if they were 40 years older than you? A different sex? From a different ethnic or social background?)

These insights and understandings are key to feminist methodology. They provide some guidance as to how we might research and understand people and their lives. They suggest to us that both the researcher and the researched are performers and that their performances must be interpreted in context (situated in time and place as well as the circumstances that prevailed at the time of the exchange). They also alert us to the operations of power and to the understanding that performances are never freely chosen by an individual. They have led feminist researchers to respect that knowledge is always partial and always 'situated' and to write this into their research. They have also led feminists to think carefully about their personal role in the research process, to consider the nature of their performances, the role of their positionality and the part played by interpersonal relations between themselves and their research subjects 'in the field' (where the research involved direct contact between researcher and research subjects). 'Reflexivity' is the term used to denote this process whereby the researcher considers her/his role in the research process and its findings. The extract below, taken from Katy Bennett's PhD thesis, provides an example of how feminist geographers write their own positionality into their research. It also shows how, by being aware, feminists might deal with the performative nature of subjectivity. Gillian Rose (1997) has discussed the difficulties involved in this process and the importance of working with, and through, uncertainty in feminist research.

Am I playing with you? *[the reader]* Maybe. But no more than the interviewee played with me. This notion of playing helps me to sensuously portray
the interview situation in which I found myself. I use the word 'found', not
just because I was physically there, but because I identified my self, my selves,
sometimes shocked my self by realising that in that interview situation there
was something rather different about my self that I rarely acknowledge, do
not often own up to. I found my self occupying positions that sometimes
were not too familiar, but with which I was still able to identify, into which I
was able to manoeuvre my self, within which I was able to locate my self,
assume that positionality. It is about being able to play the self into a
positionality, but in the act of playing, also knowing. Here I am attempting to
play at academia, playing the academic, playing with you. When interviewing
farmers' wives, I found myself playing at being the farmer's daughter,
identifying myself with the fact that I am a farmer's daughter. I found myself
playing a rather Tory role . . . Not me, definitely not me.

I speak for myself here. But by looking at identity, identifying positional-
ities, we are allowing slivers of ourselves to be viewed, the interviewee is also
going to be showing a version of self. Yes, she might be playing, viewing me
and playing, but by playing she is also acknowledging her knowing. In the
interview situation then, the interviewer and interviewee are mapping out
their intersubjective space across which they glide, come close, distance their
selves, and sometimes meet.

(Bennett 1998 pp. 128–9, emphasis in original)

THE PARTIALITY OF KNOWLEDGE

A recognition of the partiality of knowledge is not an acknowledgement of
defeat; rather, it is a demonstration of confidence. Feminist methodologies
do not shy away from the complexity of the real world, or from the task of
trying to understand it. In common with postmodernist and poststructuralist researchers, they do not hide behind a false mantle of the fictitious
'objective observer' and do not claim to discover some universal truth at
the end of the research process. Knowledge is always situated and partial.
To acknowledge this makes knowledge more human and more 'realistic'.
Feminist methodology also explicitly recognizes that the answers obtained
in research are a function of the methods used to obtain them.

Understanding that *how* we come to know fundamentally shapes *what* we
know underlines how partial knowledge inevitably is. Perhaps our feminist
methodologies will never allow us to understand more than . . . fragments of
the world . . . Because each method and each methodology provides only a
partial glimpse of the world as it turns, everyone must . . . learn to accept and

> to live with partial understandings. Why not, however, increase the chances of being surprised by new insights by seeking understanding through multiple methods? Why not combine narrative and science and geographic thinking? Feminist *geographic* methodology, especially one that incorporates multiple voices speaking from multiple locations, holds out the hope of allowing us to understand more than one fragment at a time.
>
> (Hanson 1997 p. 127)

RELATIONS OF POWER

Feminist methodology is sensitive to and explicit about power relations. Indeed, issues of power and how to deal with them have caused some feminist researchers much anguish. As feminist research is emancipatory in its goals, and as feminism is critical of unequal power relations in society, feminist research must endeavour to challenge, not reinforce, those power relations. There is widespread recognition amongst feminists that the research process is imbued with issues of power, authority, privilege, dominant voices and silences. As Susan Hanson has pointed out: 'At the core of feminist methodology is the open acknowledgement that the research born out of the research process is a joint, yet always unequal, creation of both the researcher and the research subjects' (1997 p. 122). In their choice of topic, selection of research methods and choice of questions to ask, the researcher has power to piece together a story of their choice and creation. In the 'field', researchers are often (but not always) perceived by their subjects to be knowledgeable and perhaps intimidating. In some applied research situations, the researcher has the power to affect people's lives in material ways. In the writing of their research, researchers have the privileged opportunity to speak (normally to speak for others) and be heard. In all of these ways, researchers are in a powerful position and need to search for the best ways to even out power relations, minimize researcher intrusion and enable the voices of research subjects to be heard. The selection of research methods for feminists is very much influenced by this concern about power relations.

> The politics of involvement, for feminist researchers, require research methods that recognise the relationship with others as one of (ideally) mutual concern and trust ... methods that stress mutual respect and involvement, shared responsibility, valuing difference, and non-hierarchical ways of achieving ends are not simple or shallow gestures of accommodation, nor are they just an alternative methodology. Such methods [and the way they are used] define an approach to political change. They raise the question to that of 'Who speaks with whom?' This question occurs before we enter the field

and remains with us as we engage our subjects across the space of social engagement.

(Kobayashi 1994 p. 76,
cited in Women and Geography Study Group 1997 p. 104)

As well as asking 'Who speaks with whom?', feminists ask 'Who has the right to speak for whom?' Here, feminists have engaged in a debate about insiders/outsiders (Gilbert 1994; Women and Geography Study Group 1997). Some have suggested that researching the lives of people or groups to which/whom we are 'outsiders' is at best difficult, at worst morally wrong, as it can serve to reproduce power relations. Can we as outsiders really understand others and speak for them without misrepresenting them? In answer to this, some have argued that if researchers only researched people like themselves, many people would be ignored altogether (England 1994; Staeheli and Lawson 1994; Russell 1996). Still others have pointed to the advantage of being an 'outsider' in the research context (why might it be an advantage to be an 'outsider'?). Similarly, we might argue that insider/outsider definitions are based on essentialist thought (see Chapter 6) and are thus to be treated cautiously. As Gilbert (1994) concluded, the very role of researcher and the act of writing about the project for research audiences renders the researcher an outsider to any group (except perhaps researchers themselves).

TASK

> Make a list of methods that you might employ to help you to speak to and about other people without misrepresenting them or their worlds.

FAITH IN EVERYDAY KNOWLEDGES

Research methods and ways of writing that enable, as far as possible, research subjects to speak for themselves are based upon the feminist belief that 'people know as much (probably more) about their lives, and the meanings they live with, as those who attempt to study them' (Women and Geography Study Group 1997 p. 106). This would seem to be fairly self-evident and uncontested, but much of social science methodology is based upon the underlying assumption that the researcher/scientist knows best and that knowledge is waiting to be discovered by the researcher who can decode the garbled messages of research subjects. Feminist methodology expresses faith in the value of everyday knowledges, those of the research subjects and those of the researcher, and holds the belief, as a related point, that researchers are always in 'the field'. As a result, all voices in the research process are afforded equal value. The difficulty is in finding ways to enable all to truly speak, especially within the strictures of the practice of academic writing.

TASK

- Think back to some of the research papers that you have read recently.
- Have any of the authors tried or managed to enable the research subjects to speak for themselves?
- If so, how have they tried to do this?
- If not, how might they have done so?
- Is this aim possible to achieve?

As my own research and that of others had established, in mining towns women are marginalised by many processes ... I decided to employ a number of miners' wives as co-researchers in the project in an attempt to confront some aspects of women's marginalization. The women employed had to be experiencing life with a shiftworker on a 7-day roster ... The twelve participants (three each from four different mining towns) were actively involved in the research design and questionnaire formulation and were trained as interviewers at an initial two-day workshop to conduct six recorded in-depth interviews with their friends and acquaintances. In this way it was hoped that an established rapport would exist between interviewer and respondent and would be the basis for a more relaxed and revealing interview experience ... At the second workshop held later in the year, preliminary results were analysed, qualitative results discussed and possible interventions outlined.

(Gibson-Graham 1996 p. 238; see also Gibson-Graham 1994)

RESEARCH METHODS: A DIVERSITY OF APPROACHES

Feminism does not encourage methodological elitism, but instead promotes a plurality of methods where the choice of method depends on what is appropriate, comfortable or effective.

(Women and Geography Study Group 1997 p. 98)

The quantitative/qualitative debate that has characterized modern geography (and the social sciences more generally) has also been hotly pursued in feminist geography. Since its inception, both categories of research methods have been used in feminist geography, but the bulk of researchers have tended to favour qualitative/intensive methods over quantitative/ extensive methods. Reading through the chapter to this point, it should be apparent why this is so. Simply put, intensive methods enable researchers to get 'closer' to their research subjects, to give those subjects more power to talk. It has also been suggested that it is easier to equalize power relations between the two parties using intensive rather than extensive methods, as the former allow the researched more opportunity to 'frame' (i.e. identify salient issues etc.) the research, especially where conversation-type methods are used rather than those more closely aligned to inter-rogation. More recently, however, feminist researchers have written about the potential for exploitation that arises from qualitative methods also (England 1994). Many of these methods rely on close allegiances and often friendships forming between researcher and researched. At some point the researcher needs to write up their findings, and this can cause problems where the researcher needs to write things that will upset (or perhaps even disadvantage) those research subjects (Stacey 1988; Gilbert 1994). What-ever research methods are chosen, the potential for exploitation, mis-representation and even damage is present. This is why methods need to be carefully chosen and thoughtfully executed, and why feminist methodology is so centrally concerned with power, ethics and justice.

TASK

> • In the light of the goals of the feminist project and the concerns of feminist methodology discussed above, make a list of the advantages and disadvantages of using both quantitative and qualitative meth-ods in feminist research.
> • See Women and Geography Study Group (1997) for some discussion of this.

Following the debate about quantitative versus qualitative methods in *The Professional Geographer* (1995), feminists have become more explicit about their belief that no single research method has universal advantages over any other, and that feminists should select those methods which are best suited to their particular research agendas. Often this will include using a mixture of both quantitative and qualitative methods, the former to provide information about pattern and extent, the latter to supply the contextual detail necessary for understanding. Susan Hanson (1997) makes just this appeal to feminist geographers.

CONCLUSION: FEMINIST RESEARCH IN CULTURAL GEOGRAPHY

It should be apparent to those who have read a number of other chapters in this book that there is much overlap of opinion, objectives and method between cultural, poststructural and feminist geographies, and that these approaches are not mutually exclusive. Indeed, many feminists practise what might be pigeon-holed as 'feminist poststructural cultural geography'. The intellectual heritages of these traditions are in many ways similar (for example, a challenge to research orthodoxy as a foundation), and methodologies have emerged from the cross-dialogue between them. For this reason, it is difficult to apportion credit for particular insights to particular traditions (especially as many of the key theorists belong to more than one tradition). Similarly, when reading a cultural geography research paper, it would be wrong to pigeon-hole it as 'feminist' simply because, for example, it contained considerable reflexivity. Cultural geographers too, consider reflexivity to be important. Cross-fertilization of ideas between traditions has led to the fruitful development of all, though, in this development, the contribution of feminists in challenging positivism, highlighting the gendered nature of research practice, valorizing subjectivity, emphasizing power relations and the need for reflexivity, and calling for methods that challenge existing unequal power structures has been considerable. In cultural geography, in research that seeks understandings of modes of representation and systems of cultural reproduction, many of these concerns are vital.

GLOSSARY

Positionality   A person's position *vis-à-vis* social categories and experiences, such as class, age, gender, sexuality, upbringing, nationality – indeed any factor and combination of factors that contribute to individuality.

Reflexivity   The process a researcher goes through when considering her/his personal role in influencing research outcomes.

Situated knowledge   Feminists' (and others) acknowledgement in their work, that all knowledge is social, relational and contextual, that is, situated, dependent on researcher, research subject(s), place, time and other axes.

FURTHER READING

Ewick, P. (1994) 'Integrating feminist epistemologies in undergraduate research methods', *Gender and Society*, 8: 92–108.
Jones III, J.P., Nast, H. and Roberts, S. (eds) (1997) *Thresholds in Feminist Geography: Difference, Methodology Representation*. Lanham, MD: Rowman and Littlefield.
McDowell, L. (1997) 'Women/gender/feminisms: doing feminist geography', *Journal of Geography in Higher Education*, 21: 381–400.

Raghuram, P., Madge, C. and Skelton, T. (1998) 'Feminist research methodologies and student projects in geography', *Journal of Geography in Higher Education*, 22: 35–48.

Women and Geography Study Group (1997) *Feminist Geographies: Explorations in Diversity and Difference*. Institute of British Geographers. Harlow: Addison Wesley Longman. Especially Chapter 4.

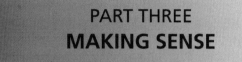

PART THREE
MAKING SENSE

Scientific Software Development's

ATLAS.ti

The Knowledge Workbench

Visual
Qualitative Data
Analysis
Management
Model Building

in
Education
Research &
Business

**New Version 4.1
Designed for
Windows 95 and
Windows NT**

Distributed by

SCOLARI
SAGE PUBLICATIONS SOFTWARE

17 Coping with archival and textual data

Kevin Hannam

This chapter considers the problem of making sense of the masses of textual material gathered from archival or other sources. It suggests ways to select, sort and analyse material so that sound arguments can be constructed. Textual data are always a representation of how people have made sense of and reflected on their own world and that of others. The term 'text' is used here in the widest sense to include other more visual cultural productions such as photographs, paintings and maps, or anything that can be read (Barnes and Duncan 1992).

When faced by texts we need to think firstly how they have been produced and secondly what they say. Some systematic way of selecting and sorting the material is needed so that it has some meaning for the researcher. This process of selection and sorting is generally subjective and creative in that researchers are actively making sense of the material based upon the knowledge they have built up doing other reading and research (Crang 1997). There is no single 'right' way to cope with and analyse material. Judgements have to be made as to what techniques suit a researcher's practical aims and theoretical inclinations, as well as the quality and quantity of data available. The aim is to interpret data and to offer a reading which is convincing but which also still offers the possibility of being read differently by others with different objectives or viewpoints. Material should not be discarded just because it seems messy or difficult to fit into a theoretical position. The experience of archival and textual work is often uncertainty and doubt; researchers sometimes have the feeling that they are barking up the wrong tree or the worry that they are distorting the truth through their selections and rejections of evidence. They need not only to have faith in their ability to make judgements about the value of materials and demonstrate relationships, but also to be willing to revise those judgements in the light of new evidence.

Once material has been located, a researcher is faced with the more difficult issue of extracting the most pertinent or valuable material for the aims of their investigation. If one is fortunate, the nature of the material itself may determine the nature of the enquiry, but one is exceptionally lucky when the facts do speak for themselves. Usually researchers have a specific question or problem in mind which has been prompted by secondary sources, and this will determine the selection of material by establishing the researcher's criteria for relevance. However, if the question is too focused, there is a danger that the researcher may impose his or her own preconceptions and that evidence may then be taken out of context and misinterpreted. In practice, therefore, most researchers will hopefully lie somewhere between these two extremes. They will have well-formulated theoretical questions in mind but will allow the material itself to suggest new, alternative lines of enquiry. At the very least researchers must be ready to modify their original questions in light of the material (Tosh 1991).

ANALYSING VISUAL DATA

Visual data proliferate in the postmodern world in which we live. Data such as paintings, photographs, maps, television or even films can also be thought of as texts that can be analysed using content, textual and discourse analysis (Barnes and Duncan 1992; Burgess and Gold 1985; Daniels and Cosgrove 1988; Aitken and Zonn 1994). Visual data need to be treated a little differently to the written text. Visual texts can, however, generally be divided into the organic (non-promotional sources) and induced (promotional sources). Indeed, promotional material possesses what Goffman (1979) calls commercial realism. This means that we are culturally disposed to view an image as real even though we know deep down that the image has been manipulated in some way. There are dangers of relying too heavily on visual sources, however. Harvey (1989b), writing from a Marxist perspective, has argued that although discourses are materially embedded, the emphasis on the study of visual images may lead to more material questions being neglected. Mulvey (1989), on the other hand, writing from a feminist perspective, has argued that visual images are part of a male, voyeuristic discourse where the men look and women are objectified and looked at. Voyeurism involves a power relation, therefore, of men over women and the preoccupation with the analysis of visual texts serves merely to reinforce dominant gender relations. Any visual analysis must, therefore, pay particular attention to the gender relations depicted in the text as well as those between the author and audience (Rose 1993). Visual imagery is, therefore, both complex and immensely powerful; however, we can outline a number of guidelines for the deconstruction of promotional images.

Analysing archival and textual materials is painstaking and generally requires a great deal of time. Few textual sources offer instant answers and certainly many are not reliable sources of 'truth', though all provide evidence. Any researcher needs a degree of scepticism when faced with any text, though even 'unreliable' texts can reveal interesting fantasies or delusions. For example, where a text is a report of something that has been seen or heard, it needs to be asked whether the writer was in a position to give a faithful account and then whether it was in his/her interest to do so.

It should also be noted to what extent there has been authorized or unauthorized censorship of the material and an attempt should be made, often by means of discourse analysis, to assess the likely intentions and prejudices of the author of the material. Indeed, sometimes it may only be material which the original author saw as incidental detail in a document that is important to a particular modern cultural researcher. The Subaltern Studies group of historians have derived most of their evidence from reading colonial texts such as court records for clues to another reality relating to imperial power. Textual research is, thus, not just about identifying a source and then exploiting it. All sources are to some degree inaccurate, incomplete, distorted or tainted; none captures the whole picture. In order to understand the significance of an event or a representation, any researcher will therefore need to examine as wide a range of sources as possible. For example, Alison Blunt (2000), in her analysis of constructions of British women and domestic defilement in the Indian 'Mutiny', uses paintings, newspapers, parliamentary debates, letters, diaries and memoirs to build her arguments.

CONTENT ANALYSIS

Content analysis is an empirical technique which involves the identification of issues and interpretation of the content of a text which is assumed to be significant (Fiske 1990; Weber 1990). For example, content analysis could be used on an archive of tourist brochures to explore different styles of representation of people and places at the same time or over time. Content analysis is a quantitative approach concerned with categorizing and counting occurrences of elements of content. The basic assumption is that there is a relationship between the frequency of a specific theme and its significance or dominance. The technique is usually used to find patterns across large amounts of material and the results are frequently presented using basic descriptive statistics such as bar charts for comparative purposes.

Doing content analysis will mean having to construct a coding frame. This is a set of themes into which material can be allocated. Possible categories could include the numbers, class, gender and ethnicity of people represented in the material. Of course, the choice of themes depends upon the questions that the researcher wishes to answer. The construction of the coding frame is done by using a logical and consistent sample of the material, say 10 per cent, and may involve several reworkings until the researcher is satisfied that everything has been covered. The coding frame should be reliable in the sense that if two people were to analyse the same material with the same aims they would come to the same results. The coding frame should also be exhaustive in the sense that all the data must be allocated to the themes, even if this means the use of an 'other' category. It is important to monitor what ends up in this category so that interesting results are not being neglected.

In her content analysis of photographs in the tourist guidebook *Lonely Planet Travel Survival Kit* for India, Bhattacharyya (1997 pp. 379–80) coded the pictures into three groups of 'authentic sights', these being 'historical sites', 'natural world' and 'social life'. She then recoded the same material according to whether people are the focus of the photographs, whether people are present but not significant, whether people are shown in a crowd, or whether people are absent. Her results indicate that over half the photos are of social life and that people are primarily the focus of this group of photos, in marked contrast to the other sights. This result should be the same if someone else analysed the photos according to the same coding frame, but, of course, different people are likely to generate different coding frames. This was a very simple example, but content analysis can also be applied to large amounts of text, in which case it is advisable to use a computer package, such as ATLAS.ti.

TEXTUAL ANALYSIS

Textual analysis is a qualitative technique concerned with unpacking the cultural meanings inherent in the material in question. The content of a text is seen as a mediator of latent and highly variable cultural themes. Textual analysis thus draws upon the researcher's own knowledge and beliefs as well as the symbolic meaning systems that they share with others. It also seeks to analyse the knowledge, beliefs and symbolic meanings being presented within the material in question. The meanings inherent in any text are, though, always highly unstable. As Gregory and Walford have argued, 'texts are not mirrors which we hold up to the world, reflecting its shapes and structures immediately and without distortion. They are, instead, creatures of our own making, though their making is not entirely of our own choosing' (1989 p. 2).

Compared with content analysis, textual analysis is richer and often more time-consuming; as a consequence, it is usually conducted on a relatively small sample of material. The results are wordier and generally use verbatim quotations of the relevant material. Again, in comparison with content analysis, it is less standardized and is often seen as less rigorous; however, its strength lies in its detailed and in-depth analysis of specific cases. Credibility of the research is maintained by the reflexivity of the researcher, i.e. the researcher needs to be constantly aware of the assumptions and preconceptions that are being made during the research and the possible impact of these on the results. Making the steps used as transparent as possible also helps in maintaining the integrity of the research and allows any conclusions to be better assessed.

Computers are undoubtedly of considerable value in the analysis of large amounts of material, but one needs to decide whether it is worth the time and effort learning how to use a program when carrying out only a small project, such as those that will be undertaken for coursework by undergraduate students. Specialist programs are used to annotate, cut and paste and retrieve information, i.e. they speed up the coding and sorting of material and help to prevent the researcher from becoming snowed under with paper. From this it becomes easier to see relationships within the material and to run speculative searches which might otherwise be too time-consuming to risk. However, one does not necessarily need a specialist program: it is relatively easy to retrieve information using conventional word-processing programs for annotation and searching. However, some of the specific qualitative data analysis programs, such as NUD.IST, can also create graphics or diagrams out of data, making it easier to visualize the relationships between codes or themes. ATLAS.ti has the advantage of being able to handle pictures, maps and music too. It is important to note, however, that computer programs will never do the analysis. The researcher ultimately has to draw out the interconnections and explanations in the material. Using a computer may speed things up and help a researcher to manage material more effectively; it may also help to reveal blind-spots and bias. For more detailed information on using computers for qualitative data analysis one should consult Dey (1993), Kelle (1995), Fielding and Lee (1991) and Weitzman and Miles (1995).

Doing textual analysis usually means going slowly and thoroughly through the material a line or sentence at a time and attempting to think about what was meant and why. The material is then open-coded, i.e. as ideas emerge they are written down alongside the text itself (ATLAS.ti is excellent for this purpose). The aim is to get as close to the material as possible in order to avoid missing anything. As the themes accumulate they will hopefully spark off more theoretical ideas which can be followed up later. The notes are then formalized into categories or codes (usually abbreviations or colours). The segments will probably end up with several codes and can be of varying lengths. Some people cut up a copy of the materials and place the coded sections into piles. (Clearly, if one uses a computer, the 'piles' are less likely to get muddled up and an individual element of text can go into more than one 'pile'.) The idea is to organize the material so that interesting relationships or themes can emerge. Using simple matrices will help. This is an iterative process in that some of the codes will break down when it is found that a particular pile of material contains significant differences and needs to be recoded in more detail. For example, whilst researching Viking heritage tourism I developed a simple code for authenticity, but then had to recode the material as I attempted to tease out distinct types of authenticity (Halewood and Hannam 2001).

DISCOURSE ANALYSIS

Textual analysis treats text as largely independent of the institutional structures which generated it and the social relations implicated in its production and consumption (Goss 1993b). However, texts are, of course, constitutive of larger, more open-ended structures that are often termed discourses. They can be thought of as frameworks for understanding and communicating. Importantly, the connotation of specific words can vary significantly between discourses. So, for example, the discourse of astronomy uses the term 'space' in a different way from the discourse of geography. Discourses constitute 'the limits within which ideas and practices are considered to be natural; that is, they set the bounds on what questions are considered relevant or even intelligible' (Barnes and Duncan 1992 p. 8). However, certain discourses will use a word in such a way that it becomes the dominant connotation whilst other discourses may use it in a subordinate or even subversive manner. Following the work of Foucault (1965; 1977), discourses have a largely unseen power because of their scientific status and the way in which this leads to the creation of specific institutions. For example, a discourse of psychiatry had the power to create certain types of institutions that we refer to as asylums.

Discourse analysis builds upon the textual analysis described above and is often referred to as deconstruction. Poststructuralist writers tend to argue that there is not just one structure behind any text being analysed, but rather is a series of often incoherent structures which are in conflict with each other.

Poststructuralist discourse analysis seeks to develop a nuanced reading that unpacks in minute detail a particular text in the cultural context in which it is embedded. Deconstruction is interested not just in what is within the text itself but also in what has been left out and the 'secret' meanings that are not obvious (Duncan and Duncan 1992). Deconstruction looks for the dichotomies that are written into any text 'and asks questions that tend to be ignored because of the dichotomy' (Feldman 1995 p. 52). It is also seen as important to note the interruptions and disruptions that occur within the smooth flow of many documents. For example, in my research into the Indian Forest Service, many original reports that I came across had pages missing, handwritten notes in the margins and scraps of paper inserted.

Discourse analysis assumes that all texts are produced intertextually in relation to other texts which are embedded within the power relations that give degrees of authority. For example, in his analysis of maps, Harley (1992) argues that these texts portray an impression of truth and impartiality which conceals their actual power of authority. Hence, a poststructuralist discourse analysis moves beyond the analysis of what is meant by a text in order to question the relative positions of the author and the reader of the text. It takes as a primary assumption that nothing is fixed and that seeking a full understanding is ultimately futile (Aitken 1997). Hence, rather than letting the words speak simply for themselves, discourse analysis treats texts as mediated cultural products which are part of wider systems of knowledge which may set the limits for, or discipline, everyday life.

TASK

> Analysing promotional images:
>
> - Select an advertisement which promotes a place known to you.
> - Try to work out as much meaning as possible, including ideas which conflict with each other.
> - Ask yourself the following context questions to help with the task:
>   What aspect of the place is being promoted?
>   What is the competition?
>   Is the advertisement part of a campaign?
>   Which medium(s) is being used for promotion?
>   What is the intended audience?
>   Is there anything missing from the material?
>   What intertextual referents are there?
> - Then examine the aesthetics:
>   What are the elements in the composition of the material?
>   What is in the foreground/background?
>   What is the location of the material (in time and space)?
>   What colours and typefaces are being used?
>   How has the material been manipulated?
>   What is the overall mood of the material?

Fowles (1996) and Williamson (1978) give further advice on the decoding of images.

CONCLUSION

As has already been stated, doing an analysis as part of a research project will usually lead to a large amount of notes which will need to be sifted through, collated and reduced in order to be able to draw out the most significant points. However, smaller projects are often undertaken as part of cultural geography optional courses and can be an interesting basis for essays and exhibitions. One should try, however, to make clear links between the aims of an investigation and the data available throughout any exercise, however large or small. At some point any researcher is likely to feel lost in the mass of data and ideas: this is normal and necessitates a change of perspective. Often with a sense of distance researchers find the important aspects re-emerge, particularly if they have someone to talk to about their work. Using a computer program may make one feel in command, but it can sometimes give the feeling of distance from the immediacy of the material; it may then be a good idea to go back to the literal handling of the raw data to renew one's enthusiasm. Finally, having completed the analysis few researchers will be able to use all of the material they have gathered; it is a little like mining for gold.

Researchers often combine different types of analysis so that the strengths of various approaches can be gained. For example, Bhattacharyya (1997 p. 380) moves from a quantitative analysis to a qualitative analysis in her study of the *Lonely Planet Travel Survival Kit* for India. She unpacks the quantitative 'social life' category to conclude that this tourist guide only deems certain aspects of Indian social life to be worthy of inclusion, 'the folk, the ethnic, the colorful, the traditional'. She backs this up with substantial verbatim quotations from the guidebook itself to demonstrate the peculiar way in which the country is packaged for tourist consumption, and this is very much in keeping with her interests in postcolonialist theory.

How ordinary readers consume texts is another important question that could be asked but which is beyond the scope of textual analysis *per se*. It should be remembered, though, that consumers of texts are not passive receivers of cultural meanings. They are actively involved in the production of meanings and are capable of the creative use of texts in ways unintended by producers (de Certeau 1984).

TASK

If you have access to the Internet, go to the archive of the American Memory program of the Library of Congress. It is entitled *Around the World in the 1890s: Photographs from the World's Transportation Commission, 1894–1896*, and consists of nearly 900 images by the American photographer William Henry Jackson. It is available at: http://memory.loc.gov/ammem/wtc/wtchome.html

- Attempt a content analysis of the photos of any one country in the archive.
- Attempt a textual analysis of four of these photos.
- Attempt a discourse analysis of the photos of any one country in the archive.
- Decide which form of analysis is the most effective for your interests (and why this is the case).

FURTHER READING

Aitken, S. (1997) 'Analysis of texts: armchair theory and couch-potato geography', in R. Flowerdew and D. Martin (eds), *Methods in Human*

Geography: a Guide for Students Doing a Research Project. Harlow: Longman.

Kelle, U. (1995) *Computer-Aided Qualitative Data Analysis.* London: Sage.

Tosh, J. (1991) *The Pursuit of History: Aims, Methods and New Directions in the Study of Modern History.* London: Longman.

Weber, R. (1990) *Basic Content Analysis.* London: Sage.

18 Handling case studies

Katy Bennett and Pamela Shurmer-Smith

Ethnographic fieldwork tries to capture significant features of the lives of the people one is working with. Whether or not one strictly adheres to the rigours of grounded theory (Glaser and Strauss 1967), whereby hypotheses are constantly tested and amended in the light of theories suggested by field observations, every ethnographer is permanently conscious that new insights are gained and old hunches abandoned as local experience deepens. As Chapters 13 and 14 have indicated, a researcher does not always have (or even want to have) a great deal of control over information gained through participant observation, interviews or focus groups. The research can rush off in unexpected directions; chance revelations can make things which formerly seemed trivial into important events; a trusted informant may suddenly be suspected of telling a pack of lies and one has to try to figure out the motive; people whom one had marked down as being friendly to each other may enter into a dispute.

No matter how painstaking one is in one's constant review and revision of notes and transcripts, the results of one's work can seem like a terrible, incoherent mess of paper when one returns to begin 'writing up'. Authoring has to take over from recording and the lives which so clearly belonged to other people now become elements in one's own fiction. Authors cannot help but be authoritarian: they impose their own order, one which makes sense to them and which they think their readers will understand. The very act of 'making sense' of the mass of (surely not senseless?) information one brings from 'the field' does violence to the realities of the people one has left there. Even researchers who had little prestige in the fieldwork situation now have the story-teller's power to select and reject from what is available; they can also slant, create heroes and victims and generally gloss in their own fashion. It is no wonder that the task is faced with trepidation.

The authoring, however, begins long before the act of crafting a text. This chapter is concerned with the stage before that of making decisions about words and style; it considers the problems of deciding what survives and what perishes and in what form. It is the stage of sorting out, thinking through and generally deciding how to use the information one has gathered from varied sources.

UNIQUE AND GENERAL

The raw material one has to work with is, in a sense, continuous with the whole of society. The act of isolating particular case studies is fairly arbitrary, as the researcher decides to focus on something manageable in order to make an explanation and then lifts it out of the mass of information. What one means by a 'case study' varies. At the largest scale, one's whole project is a case study, offered as a sample of the wider society; so, a village, club, workplace, charity, political group one has worked with may be constructed as a special case within the totality. This should not necessarily be regarded as a representative sample in the sense that any other sample would have been more or less the same; it is representative only in that, though it is unique, it is a part of a greater whole. Then there are social dramas within the limited field of the study, studies of events which can occur within the selected setting. After this, there are individual cases – people with their own particular histories, feelings and inter- pretations – who are still a part of the whole.

If, like Margaret Thatcher, one does not believe in society, then case studies will appear as nothing but random events and multiple adjacent but unique biographies. If one genuinely held to the belief that all cases are exceptional, there would be little point in making an ethnographic study in the first place, since no generalization would be possible. The ethnographer would be able to tell independent stories about individuals and then bind them together, but it would be difficult to think of a reason, other than puerile interest, for doing so. This chapter assumes that ethnographic work is undertaken not to reveal the singular, but to comprehend wider condi- tions and shared understandings.

Cases are interesting because they shed light on the context in which they occur. Every viewer of soap opera knows this; the particular illuminates the general, even if it is very aberrant. Clifford Geertz (1973) famously used a Balinese cock fight as a means of opening up the entire cultural context in which it was embedded. This was a staged event, a special and highly emotive occasion which Geertz not only was able to experience with all his senses, but could use as a way of eliciting explanations about many aspects of Balinese belief and practice. Starting from the single event, Geertz listened to multiple explanations, used these to trace through to further explanations of relationships and values, and compared these with his own observations. He generally did what any interested person would do in the course of trying to fit into a new environment. Geertz coined the term *thick description* for this multiply laminated explanation which could be gen- erated from any event.

Although Geertz is most often cited, his method could be seen as having its roots in the engaged, non-structuralist, approach of Mancunian anthro- pology. As far back as 1957 Turner introduced the notion of the *social drama*, a device whereby the origins of exceptional events (such as witch-craft accusations) were traced back over years into their roots in seemingly slight disputes. Turner's work rested on examining highly emotive situations and gaining multiple accounts of how these had evolved through shifting conflicts and alignments. In looking at exceptional cases, he elicited information on what people considered to be normal and proper, what they thought were reasonable accommodations, what they thought was intolerable behaviour. This work on the Ndembu people of northern Zambia may refer to people long dead, but it is some of the freshest, most considerate case study material ever written. Rather less personalized and intimate, Gluckman's method of *situational analysis* emerged at the same time. In *The Analysis of a Social Situation in Modern Zululand* (1958) he used a seemingly banal bridge-opening ceremony attended by local people, administrators, tribal leaders, missionaries and traders to think through the separate contexts of each of these elements and how these strands were woven together in racially divided South Africa. He pushed the analysis back into Zulu history to consider the origins of the precise moment he encountered. Gluckman was emulated by his students who used legal cases and disputes over bridewealth (van Velsen 1964; 1967) to lever open the multiple histories of the protagonists and bystanders. They took situational analysis as a starting point from which to launch into the *extended case method*, where ramifications from the initial observed situation are followed through subsequent linked cases using participant observation. One can see this as giving rise to the contemporary multisite ethnography where people are traced as their lives unfold. An example of this is given by Burawoy et al. (2000) who tracked redundant workers from a factory which had been closed down in Russia; their findings went well beyond the individual workers to demonstrate a shift to reliance on a precarious domestic economy.

The strength of all these case study methods is that they put the real lives of real people right at the centre of explanation. Their stories dictate the form of the narrative, their constructions of social relevance offer the context. Putting case studies at the centre of one's analysis allows the exceptional and the peculiar to shine through, but not eclipse, whatever passes for normality.

DISCRETION

The big problem with using real people's lives is clearly ethical. Every ethnographer needs to ask just how much can be revealed in its raw form. My own feeling (Pamela Shurmer-Smith) is that very little should be published where people can be identified as individuals. Like journalists with their convenient construct of 'the public figure', I think that people who live their lives in such a way that they court publicity do not need quite so much consideration as others, but I do believe that when a general point is being made it is not necessary to attach an identifiable person to it, just for the sake of demonstrating one's research skills.

It is not always enough simply to change people's names; sometimes it is also necessary that characters be disguised and events be shifted. Obviously the details of the literal truth are being tampered with, but the narrative is not destroyed. If this is done, readers need to know what sort of truth they are experiencing and it is probably best to employ a form, like that of the novel, which draws attention to fabrication. Because Laura Bohannan could not publish her findings on witchcraft in Nigeria, for fear of criminal action against the people she had worked with, she wrote a novel, *Return to Laughter* (1954), adopting the pseudonym Elinor Smith Bowen, so that the people concerned could not be traced through connection to her.

Similarly, Manda Cesara (1982) (whose real identity I will not reveal) published fairly salacious accounts of the sexual exploits of unnamed local officials in northern Zambia; it was culturally very interesting, but it could have led to a terrible scandal. I published nothing of my PhD research, even though there were no scandals in it, because the people concerned were very private individuals who asked me not to (and it was never going to make a novel). Similarly, in my current research I have alluded to practices in general and not given specific examples of petty corruption which would make my accounts of the Indian Administrative Service more persuasive; real people's careers would be at stake but, cynically, so too would my chance of learning more. Michael Hertzfeld (1995) similarly blurs and generalizes where he has knowledge of illicit activities in Greece. In today's climate of competitive careerism in academia, the question of revelation needs considering. If you are intending to deal with real people's lives, you will sometimes need to face up to the dilemma of whether to prove you really know what you are writing about, or to remain discreet and risk accusations of lack of evidence. Kamala Visweswaran (1994) published a fascinating paper significantly called 'Betrayal: an analysis in three acts'. She had been studying elderly women who had been part of India's freedom struggle, using a combination of interviews and participant observation. This particular story is about two women, both of whom had told her untrue accounts of their marital status but revealed information about the other. Visweswaran obtained something approaching the truth as a consequence of a very hurtful and humiliating confrontation.

The horror of my trespass lingered. I did not know how I could, or should, write about it. Indeed I thought more and more that I could not, and should not.

I recognize that the issue extends beyond my own agency and culpability: it has to do with the very organization of knowledge and structure of inquiry. Still, I want to imply neither a kind of complete, self-willed agency (I only am responsible) nor a kind of total overdetermined agency (what happened is solely the product of my training). The answer, I think, lies somewhere between the two extremes. I had witnessed one betrayal and staged another, but it was equally clear that I was a secondary character in a drama that existed before my arrival and that would continue after my departure.

(Visweswaran 1994 p. 47)

Visweswaran decided to reveal all; my feeling is that she struggled with her conscience but let her career win (and then even had the cheek to capitalize on the morality of the struggle). It would have been perfectly possible to have written a generalized account of the compromises between ideology and practice of women freedom fighters, and even to have introduced some 'it has been said . . .' comments, without personalizing. However, the fashion is for proof of close encounters.

TASK

- What is your opinion on the conflict between an ethnographer's academic obligation to reveal truth and her human duty to be considerate?
- How would you handle an interesting but hurtful revelation which would make a good case study?
- Would you behave in the same way if you did not like the people?

SELECTION

The author obviously chooses the significant case studies. These do not have to be major 'scoops'; anything which can be used as a starting point for further analysis is suitable. In a striking parallel with Visweswaran's subject, Richard Maddox (1997) worked with elderly people who had been involved in the Spanish Civil War and he became sensitive to local cultural practices relating to violence and resistance. In the manner of Gluckman, he described an incident of a bomb scare he witnessed at a bank to unpack local ideas about heroism and caution, attitudes to authority and subversive humour, but managed this without exploiting any individual person's sensibilities. He used other examples to illustrate accommodation to new mores whilst adhering to local conventions about such things as mourning and respect. These small cases can be described as *vignettes* – tiny scraps of behaviour which can enlighten an account.

Often information is generated through interviews. As Chapter 14 has shown, interviews are useful for discovering opinions and for collecting life histories; they are much less useful for revealing interaction. When case studies are based solely on interviews, this means that interviewees shift to centre stage as protagonists, whilst others are consigned to mere relational categories – 'my mother-in-law', 'my friend' – and their motives are interpreted by the teller for the purposes of her own story. Such cases may not be able to yield much more than illustrative quotations for an ethnographer's own narrative, based on other knowledge, but they can also be the starting point for gathering more information to work up into a case.

Autobiographies (where the researcher acts as interpreter and ghost-writer) make the subject position of the story-teller overt. Here the subject becomes an author and her self-centred words are, in theory at least, merely mediated by the researcher. However, even in such cases, the researcher will have selected the case, will edit it and present it in a way which is acceptable to the intended audience. There will be an introduction which glosses the story, and probably a conclusion which gives an analysis. The method of multiple story-telling can alleviate this to some extent, as in Oscar Lewis' (1964) quartet *The Children of Sanchez*, where members of a family take turns to tell their life histories in their own way, arguing with each other, showing how a family is the same but different from different sibling positions.

COSMETIC CASES

Selection of cases needs genuinely to open up areas of understanding. Mere use of local words to add 'flavour' (Cloke et al. 1997 p. 216) is an artificial device, which gives the impression of a close encounter with the people being studied, but tears their words out of context. Indeed Cloke et al. do not even tell readers the age, sex, family situation or employment of the people whose attitudes they plunder to relieve what they presumably see as the blandness of their extensive survey. A disembodied opinion is no opinion at all, and *vox pop* as a cosmetic device trivializes the important aim of trying to enable people's views to be known on their own terms.

FACING UP TO THE TASK

In this jointly written chapter sometimes it is necessary to identify a single author. From this point on 'I' will be Katy Bennett, and I shall use my attempt to transform my own field experiences into a narrative as an example of the difficulty of identifying useful case studies within one's own work.

As I pointed out in Chapter 8 on selecting a research topic, my research was motivated not just by feminist debates on theories of patriarchy but also by my own farming background and my insistence that my mother, and particularly my grandmother, were certainly not powerless within our farming family. My research on families in rural Dorset attempted to make sense of contradictions regarding the issues of power and control. I carried out participant observation in a farming household, worked as a volunteer in a local school and elicited personal accounts from a range of women by means of lengthy interviews.

When I returned to my university the resulting stack of notes, diaries and transcripts offered a daunting prospect. The fieldwork had seemed so real and immediate, but the 'data' now looked alien. There are plenty of texts which reveal the experiences of other researchers on methodological issues in relation to doing fieldwork, but there is relative silence on the handling of case studies when one returns. Whilst the debate on the writing of research has relevance here, with its attention to the extent to which ethnography is 'something made or fashioned' (Clifford 1986), there is still a fudging of methodological concerns between the doing of fieldwork and the process of writing.

When it comes to making sense of notes and transcripts, researchers turn to guides on how to analyse their materials. There is good advice on the process of coding, managing and analysing data (Crang 1997) but, for me at least, this is one step too far ahead when I face the disorganized fragments that make up my study. My initial concern is what I should look for in my data, and later there is a lack of confidence in knowing whether or not segments of text are worthy of a particular code. The crux of the matter is my concern that I make good use of my material and do not miss potentially valuable insights. Also, as indicated by the grounded theory approach, I worry about being flexible enough to allow issues to 'emerge' from my material and hope that I am not too blinkered by my particular agenda and perspective.

Whilst reading for help on how to handle their data, researchers are inevitably introduced to software packages that purport to aid analysis. There are a range of these available, with each one having different advantages and disadvantages depending upon the needs of the researcher. Their seductive claims, especially when they promise to assist a grounded theory approach, make them very enticing, but they cannot really generate theory.

To help me handle my transcripts and notes from my research on farmers' wives I chose to use NUD.IST (Non-numerical Unstructured Data Indexing, Searching and Theorizing), a multifunctional software system for the development, support and management of qualitative data analyses. Conceptually, NUD.IST consists of two databases, the first being the document system containing the data themselves, and the second being the index system containing information about the data. I chose the package because of these distinct and separate databases, with my analysis kept separately from my transcripts, so that when I needed to shift and change my analysis, I was able to return to my bare 'data'. I also decided to use the package because of its flexible index system, which could easily be shifted and changed to reflect changes in my analysis.

Whilst NUD.IST certainly helped me with the management and organization of my research material, it actually hindered my analysis. Before too long, I realized that I was thinking within the capabilities of NUD.IST, allowing the package to structure my analysis in a way that was not beneficial to my research. In particular, I was fiddling with small pools of my data and losing sight of the complexity of their whole context.

Researchers need to be very clear, before they start, how they want to use software packages to help them with the analysis of their data, otherwise the package can come to dominate. When I realized that I had lost sight of the direction of my analysis, I returned to hard copies of my diaries, notes and transcripts, reading them through and re-engaging with their complexity. I wrote notes down their margins and began to use coloured pens to flag up issues and themes, isolating what could be seen as small cases and vignettes which seemed to illuminate my concerns about women and power. There were times when I kept getting lost and was unsure as to what to make of particular parts of conversations. Time and again I had to pull away from the data to reconsider the initial aims of my research, my theoretical motivations. In a way there was a certain amount of swinging between getting lost in the data and trying to distance myself from it in an attempt to reconfigure the picture of my research. The research experience can be compared to looking at a painting: first standing up close to it and getting lost in the brush strokes and unexpected surprises, such as the colours used to depict skin tone and then stepping away from it to see the picture in its entirety.

As I repeatedly reread my transcripts and notes, it became evident that there were issues to which farmers' wives kept returning that related to my concerns about patriarchy and autonomy. These revolved around kinship and consumption matters, such as shopping rituals, the politics of meal organization for the wider family, and conflicts over household bills with mothers-in-law responsible for their payment. Up close there was just detail, but themes began to emerge which addressed my theoretical motivations regarding patriarchy and the control that women, particularly mothers-in-law, had over kinship relations through consumption practices. This had repercussions for the survival of the farm business and, more importantly, the farming family. I decided to focus on a case study of one family's fairly idiosyncratic meals to open up a whole range of other provisioning practices within the community.

Central to the analysis of case studies is thoroughly knowing all the research material and moving between its detail and the objectives of the research, using the two to impact upon each other. It is not an easy task, and hence there is so little discussion of it in ethnographic accounts. The movement from feeling lost to crafting a satisfying account has repercussions for critical readings of ethnographies, with the need to give plenty of consideration to why authors focus on the particular incidents they do and how they present these to tell a 'story'.

When themes begin to emerge that inform the theoretical aims of the study, researchers attempt to relate them to each other and to collate pools of material that deal with particular issues. These are then worked into chapters that make an ethnography. This is not a simple process but often requires considerable rewrites as words and chapters are massaged into place. With me, initial drafts of chapters are deluged with detail, so that central issues are not always apparent. With rewritings I edit and reduce the detail to strengthen the central issues of my research by focusing on incidents and case studies. Maybe it is just me, but with these rewritings I write myself further and further away from original field notes and I worry that I am turning my back on the voices of the people being researched as my interpretation comes to the foreground. Writing case studies, though, is not about the truth in the singular, but about partial truths and multiple understandings.

As well as note-taking and recording interviews with a Dictaphone, I used video during my fieldwork. Everyone knows, but it is easy to forget, that making a video recording provides only a partial perspective; a camera is set to record an event, but the perspective of the ethnographer frames what is recorded and then becomes the case study. The camera is positioned and shots are selected which are later edited to tell a story fashioned by the film-maker. The people being recorded know what is happening and are affected by the presence of the camera; they behave unusually or differently, but within their repertoire of behaviours. It is easiest to use a video camera when the situation is seen as something special: I recorded a fund raising event and a children's party at considerable length without anyone considering it to be unusual behaviour, but it would have been impossible to record a family quarrel.

I needed to make sense of case studies emerging from the stories I had been told in interview situations and then to splice these into accounts from other members of the same extended families interviewed separately. A picture was forming in my mind of a particular group of women related by marriage, who seemed to illuminate many of my ideas about individuality within patriarchy. I felt that I could understand their sentiments and their interactions even though each had her own perspective and they frequently harboured resentments of each other. Instead of trying to establish a single representation I decided to tell the story of these women as a play, with the acts set in the homes of the different women. It was, of course, an exercise in imagination – I could not know how the women behaved when they were on their own or how they talked inside their own heads – but by shifting my interpretation into a drama, I felt that the element of my own fiction could be more truthfully represented.

It is important when reading ethnographic accounts to see case studies for what they are – events which ethnographers have decided to pull out of the mass of data in order to hang an explanation upon. Sometimes these are trivial, like Anthony Cohen's twice-told tale of 'Ertie's Greatcoat' which illustrates Shetland humour, friendship and dress symbolism and allegedly gains local colour from being written in dense dialect (Cohen 1986; 1987). Sometimes they are deadly serious, like Feldman's (1991) constructions of violence and protest in Northern Ireland. Always the intention should be to go beyond the anecdotal to reveal the sensitivities and knowledge construction of a community, to show the cultural formations in action. If case studies do not do this, then they are no better than gossip journalism.

We have referred throughout this chaper to the research situation, but case studies can also be used in essays, fieldwork presentations and other small projects. They can be the means whereby a local encounter can be used as a device with which to piece together the fragments of background knowledge and observation. A good case study makes many things seem suddenly to slip into place, but one should not be fooled into thinking that case studies are easy to generate, just because they read easily.

GLOSSARY

Extended case method A way of studying a community through time through a series of linked case studies.

Situational analysis Starting one's interpretation from a situated and bounded event.

Social drama An unfolding occurrence which members of the community consider to be noteworthy. A dispute would be a good example.

Thick description Milking every possible interpretation out of a situation by soliciting explanations from many people and then asking for explanations of explanations.

Vignette Small illustration.

FURTHER READING

Mostly the reading for this chapter needs to be extracted from ethnographic accounts. It is anthropologists who have developed the case study method but some geographers who have made good use of case studies are:

Keith, M. (1995) 'Ethnic entrepreneurs and street rebels: looking inside the inner city', in S. Pile and N. Thrift (eds), *Mapping the Subject: Geographies of Cultural Transformation*. London: Routledge. pp. 355–70.

McDowell, L. (1997) *Capital Culture: Gender at Work in the City*. Oxford: Blackwell.

Waterman, S. (1998) 'Place, culture and identity: summer music in Upper Galilee', *Transactions of the Institute of British Geographers*, 23(2): 253–67.

From an anthropologist, useful background reading would be:

Geertz, C. (1988) *Works and Lives: the Anthropologist as Author*. Cambridge: Polity.

19 Representation of research: creating a text

Katy Bennett and Pamela Shurmer-Smith

For too much of the time learning is a sponge-like activity, soaking up other people's material and then squeezing out the same stuff. Thirty years ago C. Wright Mills (1970) urged young sociologists to write something, anything, every day, not just to record facts or to exercise their writing skills, but to develop what he called a 'sociological imagination'. This use of 'imagination' has gained currency because it captures so well the twin aspects of what any creative person tries to achieve: an image of something which already exists in the world and the conception (in both senses) of something new. Cocteau (1947), playing with two words for 'writing', maintained that, 'Écrire est un act d'amour. S'il ne l'est pas, il n'est que l'écriture' ('Writing is an act of love. If it is not, it is just recording'). Imagination, then, is a matter of involvement, where the creator of a text cannot stand back and claim only to be a clerk keeping the records. Octavio Paz ends his poem 'A draft of words' with 'I am the shadow my words cast' (1997 p. 155). The text makes its author just as much as the author makes the text, and for this reason presenting one's work can be a terrifying (and exhilarating) experience. Whether it is a small piece of course work or a major piece of research, creating a text should always be a labour of love.

> No sooner I write . . . it is
>   Not true
> And yet I write hanging on to
>   Truth.
>
> *Note:* 'I' refers to Hélène Cixous.
>
> (Cixous and Calle-Gruber 1997 p. 10)

The phrase 'writing up', applied to a piece of research, always sounds so depressing, perhaps because it is so close to 'wrapping up', with its double meaning of finishing and concealing. The writing-up stage is so often a time of disillusionment, brought on by the realization that the exciting contribution one was intending to make to the state of knowledge can never be anything more than a dull report – mere *écriture*. It is sometimes also a time of shame that one has accomplished so little and guilt that this means that one has to compromise by writing 'up', padding with words and bolstering with references. The passion Cocteau invokes cannot admit this sell-out.

THE CRISIS OF REPRESENTATION

Despite their best intentions, however, it is a common experience for people (undergraduates tackling a course assignment or their dissertation, PhD students writing a thesis, postdoctoral researchers crafting a monograph or a paper for a journal) to find themselves confronting a mass of data which seems unwilling to resolve itself into a text. Consciousness of the 'new' ethnography of the 1980s (Clifford and Marcus 1986; Marcus and Fischer 1986; Strathern 1987) only made the task of writing cultural accounts more difficult, since it placed an emphasis on ethnography as authored text. This meant the recognition that the author was embodied and subjective, rather than disembodied and objective. Suddenly the ethnographer was *there* in the writing; no more 'It was observed that . . .'. When the author rejected the passive voice for the active, use of the first person ('I') became necessary; there could be no further pretence that anyone would have interpreted the same way. At the same time the people being researched were not so easily seen as the 'objects' of a study but were rather recognized as cooperating in it. Today it not only seems pretentious to write about cultural phenomena in a 'scientific' register, it also seems dishonest, for a cultural observer cannot avoid being to some extent a participant (Srinivas 1979). But it is still difficult to know how to deal with these issues and write an interesting text at the same time.

The expression 'crisis of representation' was coined by the anthropologists George Marcus and Michael Fischer (1986) to communicate the problem that was experienced by ethnographers who realized that, when they tried to write about the world they observed, they could not avoid being taken over by the political assumptions inherent in the modes of representation available to them. Sensitive ethnographers realized that in writing about people they were, in a sense, creating them in the form in which they were written. They also realized that they were writing themselves into a tacit acceptance of a mode of communication which gave the primary voice to the author.

The only way to an accurate view and confident knowledge of the world is through a sophisticated epistemology that takes full account of intractable contradiction, paradox, irony, and uncertainty in the explanation of human activities. This seems to be the spirit of the developing responses across disciplines to what we describe as a contemporary crisis of representation.

(Marcus and Fischer 1986 pp. 14–15)

Practically everyone who has seen a representation of themselves, a place they know or an event they experienced feels wronged, dragged out of context, misunderstood (Hastrup 1992). Somehow, it always seems as if some essential truth is missing from any representation of something one knows intimately. The crisis of representation emerges from the (impossible?) desire to represent without doing violence, when the characteristic forms of expression of people being represented may not resemble those of Western science or art. Increasingly cultural geographers are looking to a range of texts not just to find out how places are, but to try to learn how and why different people represent them the way that they do.

The new ethnography has required that writing become dialogic. It needs to enter into some sort of dialogue with the people who are being written about, they need to be able to recognize themselves in the text and feel that they have helped to create it, but, as Clifford wisely indicates, this 'is not to say that [the] textual form should be that of a literal dialogue' (1983 p. 135). Thus the incorporation of the words of the people who have helped to create the text is not inevitably an exercise in sharing; in the end, their words are invariably appropriated for the writer's benefit and the writer crafts the best text she can.

Quotations are always staged by the quoter . . . a more radical polyphony would only displace the author, still confirming the final, virtuoso orchestration by a single author of all the discourses in his or her text.

(Clifford 1983 p. 139)

Following on from this recognition of the author's subjective voice, it has become crucial that readers can tell where an author is coming from; issues of positionality and emphasis on the way in which information was gathered are now an essential part of any new text. In the hands of self-centred people this degenerates into an excuse for auto-ethnography (writing about that fascinating person 'me'). Many detractors cite this as a weakness of the new ethnography, but it should also be regarded as a strength, since the new textual conventions help readers to spot authors who depict themselves as the heroes of the text and unwittingly show how limited the authorial gaze has been. So, authors need to make it clear how they came to the writing (Cixous 1991), but unless, like Cixous, their own writing is the subject matter, this should not drown out what is being written.

In theory, much of the problem of how to present work disappears if one follows Mills' advice and writes all the time, not just records, but also interpretations which are genuinely self-reflexive and always conscious of the growth and change through the previous stages of one's research. Whilst this is a reasonable expectation when performing an analysis of texts, it is an almost impossible ideal when doing ethnographic work, simply because it is difficult to find the energy. However, impossible or not, it is certainly worth trying to keep the question of representation in mind all the time when doing ethnographic fieldwork. The point is to try never to divorce the two components of doing cultural work – the understanding and the creating. If this is achieved, one writes rather than 'writes up'.

Landscape of words:
my eyes, reading, depopulate them.
It doesn't matter: my ears propagate them.
They breed there, in the indecisive
zones of language, the villages in the marsh.
They are amphibious creature, they are words.
They pass from one element to another,
they bathe in fire, rest in the air.
They are from the other side.
I don't hear them: what do they say?
They don't say: they talk and talk.

(from 'A draft of shadows', Octavio Paz 1997 p. 145)

FORM

Not all representative work is about ethnographic research, however. People have to create texts throughout their university careers and one does not always have much say in the selection of form of representation of one's work. University lecturers, for example, have the habit of setting assignments which consist of essays with word limits. Given the general demise of the essay outside academia, it is reasonable to question why a form which was popular with the Victorian middle classes has such an entrenched position inside. It is worth finding out whether there is a good reason for the conventional essay (where close attention has to be paid to the exact title) or whether something where the author has more control would be acceptable. In a field such as cultural geography it is appropriate that up-to-date styles of communication are practised as well as providing objects of study; this is not just for effect, it is a matter of being in keeping with the theoretical position and subject matter.

Independent studies, where a student generates a topic of research with the guidance of a supervisor, are a fairly routine part of final year undergraduate assessment in universities world-wide. Geography departments in British universities are becoming increasingly flexible in their regulations for the independent study, though the dissertation, with a word limit of about 10,000, remains popular, perhaps only because it is customary. Just as it is worth enquiring whether essays really are the only acceptable form of coursework, it is also sensible to explore the possibility of other types of independent study presentation, especially if one has talents in other directions.

Down with posters

Whatever the scale of the task, ranging from an unassessed class based activity upwards, it makes sense to select a form of representation which best conveys one's ideas. Although posters have been fashionable for some time, they are frequently disappointing, unless they form a sequence or are backed up by an exhibition or a verbal presentation. If one thinks of the uses to which posters are put outside academic contexts, one can see why they are rarely satisfactory for academic work which rests on marrying theoretical and empirical concerns. Posters are intended to grab the attention and convey an immediate, unambiguous message; they cannot be pored over and read slowly. We have seen far too many posters, both as undergraduate assignments and as displays of research at professional conferences, which are really just essays on a poster-sized card – too many words for the medium, too dense, no one can follow a complex argument standing looking at a wall. It may be that the only valid reason to make a poster is to draw attention to something else one has created – an exhibition, a video or whatever.

Photography and video

Photography and video are valuable textual devices, but in an academic context the images they generate cannot be left simply to 'speak for themselves'. They need bracketing in some way, so that the authorial voice is made clear; the author needs to be seen to be making a theoretical point, not simply showing pictures. This means that photographs have to be accompanied by written or spoken text, whether in a folio or an exhibition. Similarly, video requires commentary and is usually improved by the use of music to create atmosphere.

Obviously, to make an effective video one needs access to editing facilities; these are becoming more available in Western universities as expensive dedicated editing suites are giving way to computer editing.

Video potentially offers one of the best opportunities for presenting the words of other people, and is certainly a good way of handling dialogue and interviews. However, the end product is always used to convey the message of the author of the video text and, in the interests of honesty, it is important that the selection and editing are made overt. The viewer needs to be made aware that, however unobtrusive the video-maker, it is authors who allow the words to be heard and who place them in their own preferred context.

The script

Writing for a video needs considerable thought and takes time. People often rush into video-making, get carried away with shooting, selecting images, editing and so on, and then add a commentary. A shooting script should exist before a video is shot. This roughs out the story that the images should tell; it looks a bit like a comic strip, as you sketch out the shots you need to make a contribution to the narrative, and some of the words are likely to come through into the final commentary. (You can buy special stationery, with the space for sketches and words, but this is mainly for effect.) The script might change as images play their role in a production, forcing the commentary to shift to fit and explain them. The script locates the author (the video-maker) and might be a voice, voices, music, written text, a combination of these and so on. It might talk directly into the images, drawing attention to key moments and explaining them, or it might explain in a less direct way, forcing the audience to play a part in sense-making. When it comes to editing, one gets a smoother version if the images are fitted to a sound track which has been laid down first.

There are many ways in which videos can be explained and there is plenty of scope to be creative. Make sure, though, that attempts at creativity do not mask messages and purposes. A good way of ensuring that this does not happen is to show work (and accompanying text) to friends for feedback on what they did or did not 'see'. Although it is important to explain images captured by video recordings, it is impossible to justify all. The trick is to make clear the crux of an argument and to guide the audience through your way of seeing and explaining.

Multimedia

Written texts, maps, photographs, video, music can all be combined to make CD-ROMs or interactive websites, allowing the audience to navigate text and images in a different way than is allowed by hard form. The possibilities are enormous, with the audience dipping in and out of video and text, escaping the linearity of words alone and traditional methods of reading. Sometimes it is valid to make use of the power the author grants to the audience with interactive media, to allow conscious decision-making in the interpretation of the text which is being presented. Producing CD-ROMS and websites is nothing like as difficult as it looks and the learning curve is a very steep one (with liberal help from sympathetic technicians).

Exhibitions

Being exceedingly wary of the need to avoid constructing something that harks back to the primary school 'nature table', exhibitions of artifacts can be an effective way of communicating some cultural geography themes. The three dimensional is often more impressive than a flat display, as is the use of texture. Although the most obvious use of exhibitions is to provide exemplification of material culture, they can also be employed to make abstractions concrete. Advice on display techniques can usefully be gleaned from fellow students in art, architecture, design and media departments, but a great deal can be picked up from looking at museums, promotional exhibitions and even shops.

An effective way of acknowledging the perspective of the author of an exhibition, video work or a CD-ROM is to accompany the project with a leaflet, probably a folded single sheet of paper, which sets out its intentions, puts it in context and offers a bibliography.

When they are well done, alternative forms of text can make essays and dissertations pale into insignificance, not just because they are more attractive (though this is undoubtedly the case) but also because they can say so much more and say it more subtly, catching multiple meanings, allowing a foregrounded message to be expressed at literally the same time as the background is displayed. When they are badly done, they are no worse than bad essays, though they can be embarrassingly public.

EXPRESSION

As Mills suggested, presenting findings is not the sole purpose of creating a text. The activity forces one to be conscious of how texts work by judicious selection, editing and positioning. In the words of Matisse, 'The simple things are the most difficult to express', and forcing oneself to express creatively is the best possible way of really understanding whatever one is trying to communicate. Since many of the methods referred to require group work, useful discussions inevitably take place about theory, interpretation of observations, effective ways of communicating. People surprise themselves by arguing passionately about subjects that, in a seminar, it would be difficult to get them interested in. They also find themselves willingly doing far more work than usually goes into the production of an essay.

PUBLISHING

There is certainly nothing wrong with continuous prose on paper, but it is sad that so many essays are read once by the marker and then are no benefit to anyone. Writing becomes more attractive when it is published and distributed, even if only at the most immediate level. We have seen interesting results from producing journals on themed topics. Sometimes these emulate academic journals, with a selection of serious research papers followed by book reviews. At other times they take the form of special interest magazines: in this case appropriate layout, attention to typography, selection of photographic images form a significant part of the exercise. Either way, individual authorship is combined with group editing and refereeing, which can be painful but is very effective in raising both critical powers and the standard of finished work.

These do not have to be printed. They can be published online and become a valuable resource for students in the following years (with all the usual warnings about plagiarism, which is such a mean-spirited, uncreative sort of crime that it constitutes its own punishment).

ALTERNATIVE WRITING STRATEGIES

Here there is a difference of opinion between us. Although she revels in the use of new representational strategies, one of us abhors attempts at alternative writing. She sees poetry, mini-plays (Jensen and Reichert 1994) and meaning-free words as inappropriate to academic writing. She does not know whether she is supposed to take Allan Pred's (1994; 1997) 'poetry' seriously as poetry, and if it is not poetry, she doesn't know what it is (or much care). After an initial appreciation of the joke, she has wearied of getting essays where some of the pages are printed on acetate sheets (to make the layers of thought intrude on each other) or where the pages are arranged so that the work reads from back to front (to emulate the experience of drawing into a station with one's back to the engine). Such devices irritate when one realizes that the medium is probably going to be the whole message. She cringes at the thought that one day she might have to witness geographers at conferences indulging in 'all of those techniques which work with bodies – various forms of dance and music therapy, contact improvisation and so on' advocated by Nigel Thrift (2000) as the next wave of representation in cultural geography. She knows, however, that this is a personal prejudice and listens to the other, who sees these techniques as offering a way forward into a new, more engaged form of academic work which harnesses all of the senses.

The other of us revels in the possibilities of alternative writing and is interested by the work of some sociologists who explore its possibilities for qualitative research (Richardson 1994; Ellis and Flaherty 1992; Ellis and Bochner 1996). Whilst understanding the other's hesitations and herself not grasping how certain ideas for performances might be realized, she enjoys writing and presentations that attempt to escape geography text at its most humdrum. As an undergraduate at Portsmouth, in her cultural geography she learnt in a different way to think more freely and was encouraged to express herself in sometimes unusual, more meaningful, ways. She found this a revelation and for the first time felt almost capable and much happier. What she enjoys about alternative writing is that words are often written to be performed, to be read aloud.

Alternative writing seems to be more overt in its grappling with language and choosing of words to create text. Whilst all texts are creative, the creativity of alternative writing is more apparent, with authors purposely massaging words to tell their story of their research or whatever. Using words from transcripts and notebooks to write a story, a play, a piece of poetry might better evoke the issues that an author is attempting to portray. Furthermore, the context of the research situation might be more incorporated into the text of alternative writing, with, for example, the heated crossfire of parish council meetings being written through in the form of a play. Maybe alternative writing is more up-front in its 'use' of its research material, quite openly manipulating 'data' to make a point. Too often, the (ab)use of the people who are being researched is less obvious in dissertations, theses and papers which drag in quotes from their interview transcripts to support the argument that they are spinning.

Okay, of course there *are* plenty of excerpts from transcripts in texts. But they are content driven quotes. Which comes out of the whole *coding process*. Hhhh, after all, this is the way I've worked as well, sifting through transcript after transcript trying to code them, to crunch them down. It's all for a good reason since, once they're coded, then I can extract all the relevant material on *the* body or *the* private sphere or anxiety and work or, or . . . So, coding has made me do some laborious interpretative work and, most painful of all, I've summarized hundreds of pages of speech into key codes. But then, I just cut and paste a nice quote as a lone representative voice into my text, turning once spoken, non-written, words into *what* they as representatives (sometimes of discourses or sometimes of social groups) said, not how they said it and in what situation. Quotes that are all too completely shifted into written language, rather than shifting written language toward spoken language, and which hardly ever refer to the dialogue that they were found in. If you think that one of the fundamental principles of conversation is turn-taking (i.e. one person speaks then another person speaks) this principle disappears in most texts that claim to *represent* voices that *represent* discourse. And, so, yes, dialogue needs to be there too, or perhaps talk, since

> it then loses its associations with a lot of the textual emphasis of dialogical theory.
>
> (Laurier 1999 p. 38)

It is certainly the case that those doing cultural geography are not usually poets, novelists, playwrights (film-makers or artists) but this need not be so important. What is imperative is the message, the research, which is being conveyed. If poetry or a play provides appropriate support for the writing through of research, then maybe there should be space provided to allow for this. Those acetates referred to earlier got in the way of the message. In this author's thesis, which focused on issues of patriarchy and domesticity in family farming in Dorset, she wrote a 'story' chapter that attempted to capture the complexity of farmers' wives lives in relation to consumption and kinship, against which more explicitly analytical chapters would later sit. This chapter focused on the lives of four women who either had married into or were part of a particular farming family, and was based on one fictional day. The story slid between the different and complex perspectives of the women in relation to the farm, their positions within it and relative to one another, and their conflicts over issues of consumption. It attempted to show not only differences and tensions but also the extent to which they were sewn into each other's lives. The chapter was set up as a piece of 'fiction' but was based on observations and interpretations of their stories.

READERSHIP

We have been considering ways of creating texts which might best convey the theoretical implications of one's work. It is important always to consider who constitutes the anticipated readership and the skills they will bring to a reading. The notion of appropriateness needs always to be present, unless one is producing text only for oneself or people one is familiar with. Much experimental work is for fun and it takes risks. It can help one to sort through one's ideas and, importantly, to distinguish theoretically founded positions; it is a good way of learning, but not necessarily a good way of communicating. Indeed it often makes a virtue of indeterminacy, but at some point one needs to take the responsibility to clarify what one knows. As with the work described above, to be acceptable in an academic context alternative writing needs always to be framed in more conventional explanation.

In the case of funded research, texts must respect the stipulations of commissioning agencies which specify their requirements; these are often policy-makers. In these situations agencies play an active role in the formation of texts, often giving guidelines on the sorts of images and styles that reports must adopt. This can be illuminating, challenging and frustrating, not just because one feels that one's creativity is being curtailed, but because even conventional academic usages are cut out as 'jargon'. Facts are required, even where facts are not the best way of communicating what one believes to be the truth. In reports, complexity must be simplified and findings smoothed more than many cultural geographers would wish. Judith Okely (1983) has written of the frustrations of doing work with gypsies on behalf of the Home Office and feeling uneasy about the 'ownership' of the end product. However, sometimes this rigid cycle of knowledge creation can be subtly transgressed by incorporating, for example, local poetry and photographs to portray the special qualities of a place. On the whole, though, people need to think before accepting a commission which will require them to produce representations which they find uncongenial.

CONCLUSION

This chapter throws up some challenges to those doing cultural geography. Although it contains some differences of opinion, what is clear is that conventional ways of creating texts in universities should not always be taken for granted, but sometimes should be questioned and challenged if more meaningful media can be appropriated to convey messages and research. Challenges to privileged knowledge and ways of knowing, such as those instigated by non-Eurocentric, non-elite, feminist and poststructuralist concerns, might be better supported by different ways of creating texts.

FURTHER READING

Ellis, C. and Bochner, A. (eds) (1996) *Composing Ethnography: Alternative Forms of Qualitative Writing*. London: Sage.

Laurier, E. (1999) 'Geographies of talk: "Max left a message for you" ', *Area* 31(1): 36–45.

Rabinow, P. (1996) 'Representations are social facts', in *Essays on the Anthropology of Reason*. Princeton: Princeton University Press. pp. 28–48.

Richardson, L. (1994) 'Writing: a method of inquiry', in N. Denzin and Y. Lincoln (eds), *Handbook of Qualitative Research*. London: Sage.

# 20 Concluding thoughts

Pamela Shurmer-Smith

At the end of writing a book it is almost inevitable that one's thoughts turn to what was not written. Writing is a cultural activity; when the writing is aimed at students who study culture, reflecting on the process of construction implies getting into a spiral of self-referentiality which could exist only in the strange cultural niche known as academia.

One of the problems we all faced as we wrote was that we wanted to write about the exciting things we were finding out in the field of cultural geography and we had to keep reminding ourselves that this was not what we had set out to do. We have probably expounded more than we originally intended. This is a major problem with people who earn their living as teachers: we can't resist the urge to pontificate. We wanted a book which was made up of opportunities for students to think their own way through the subject; that is why we called it *Doing Cultural Geography*. We wanted there to be lots of lively, short extracts from recent and classic publications which could stand on their own but also which would act as samplers of the whole chapter or book. We hoped that these would combine with suggestions from us to prompt new work. In trying to achieve this we made some strange discoveries about the field of study that we thought we knew.

It took each of us a while to confide in the others, but we gradually realized that an awful lot of published cultural geography is very derivative. We all found it quite difficult to find enough interesting passages of text for students to work on that were not studded with references. 'One small idea and a lot of padding per publication', is someone's exasperated comment. We tried editing out the references and saw the texts fall apart before our eyes. We assume that this situation is the direct outcome of the cultural practices fostered by the research assessment exercise. Unless one hurries, the published return on a piece of original research (particularly research based on lengthy ethnographic methods) is liable to fall outside the assessment period in which it was carried out. Just as the limitations of early recording techniques meant that music had to be produced in very short units, necessitating changes in the form of the popular song, so the need for rapid research publication is changing academic output.

We'd been experiencing the change in published work from so close up that we hadn't appreciated its full extent. The audit culture, intimately linked to strategies of minimizing risk, has gained a hold in academia, just as it has in most aspects of late modern life. It creates its own value systems and it influences the construction of new knowledge. This is hardly surprising: metaphors relating to industry regulate industrial society, metaphors relating to accountancy regulate post-industrial society. That's just the way culture is.

Related to all this, we realized was that the sort of stuff we were accustomed to reading could seem almost unintelligible if one had not long been a part of the discourse. We had also learned to read as if on a raiding party, grabbing what we needed to keep up to date but hardly ever savouring the construction of the text. Writing this book made us start reading like our students and suddenly we realized why they had so many problems with the things we had been referring them to. We were rather ashamed to make these discoveries since we'd always rather prided ourselves on being close to our students.

The writing team emerges from the Department of Geography at the University of Portsmouth and the book came out of an idea floated (in the pub) after a successful PhD viva. Katy, Tim and Kevin had all been undergraduate and postgraduate students together, I was their supervisor, and we realized that our close group was disintegrating. We felt that we had a special take on cultural geography and we wanted to capture this in a piece of joint work. Carol was a relatively new lecturer in the department, but we recognized a kindred spirit and asked her to join in.

As we started writing, Kevin was teaching at the University of Sunderland, Katy was working on a research project at the University of Durham, Tim had become a member of the Geography of Health team in Portsmouth, Carol had just started her research on freelance workers in journalism and I was becoming increasingly involved with work in India. These changes were both a strength and a weakness. Open to new influences, each of us was developing in different directions which meant that we became more varied in our approaches, but it also meant that we disagreed with each other more often. Suddenly we were no longer a mutual admiration society but a collection of individuals who had once worked together. The younger members of the team had the added burden of the insecure employment which plagues academia; there were also babies and marriages which made writing difficult to concentrate on. The permanent dialogue we had anticipated at the outset of the project did not materialize. Except where there is joint authorship the chapters are very much more monologic than we had hoped. In a way we think that this, too, reflects the current state of academia. But we have tried; this expression of tristesse is probably just postpartum depression.

What we hope we have been able to do is to encourage students to claim their discipline back for themselves. We want people to feel that cultural concerns are implicated in all aspects of geography, but we certainly do not want to make any prior claims for something called culture. We hope that the book has stimulated a view that culture is what one does, not what one has, and that this doing is infinitely subject to change, whether in innovation, imitation or reaction.

We want to emphasize that the manner in which things are thought and done is important, sometimes even of deadly significance. Although many students are attracted to cultural geography because it seems to be a warm, humane field of study, cultural practice can turn cold, cruel and exclusionary. So long as there are barriers to the construction of common meanings, the concept of the alien will exist. This does not imply that people all have to become the same, just that they need to have the knowledge resources to be able to comprehend each other. Of course this is an ideal, but academic pursuits are supposed to embody ideals.

We could have structured the book along lines that offered advice about discovering cultural practice in relation to the major concerns of economic, political, social, environmental and development geography. Certainly these issues have been in our minds as we have written and we would hope that they have been in our readers' minds too, because it has never been worthwhile to think about geography or culture without including everything that human beings do.

References

Ahmad, A. (1992) *In Theory: Classes, Nations, Literatures.* London: Verso.

Ahmad, A. (1995) 'Postcolonialism: what's in a name?', in R. de la Campa, E. Kaplan and M. Sprinkler (eds), *Late Imperial Culture.* London. Verso.

Ahmad, A. (1996) *Lineages of the Present: Political Essays.* New Delhi: Tulika.

Ahmad, A. (1997a) 'Cultures in conflict', *Frontline*, 14 (16): 76–8.

Ahmad, A. (1997b) 'Reading Arundhati Roy politically', *Frontline*, 14 (15): 103–8.

Aitken, S. (1997) 'Analysis of texts: armchair theory and couch-potato geography', in R. Flowerdew and D. Martin (eds), *Methods in Human Geography: a Guide for Students Doing a Research Project.* Harlow: Longman.

Aitken, S. and Zonn, L. (eds) (1994) *Place, Power, Situation and Spectacle: a Geography of Film.* Lanham, MD: Rowman and Littlefield.

Akhmatova, A. (1976) *Selected Poems*, tr. D.M. Thomas. Harmondsworth: Penguin.

Anderson, B. (1990) *Imagined Communities.* London: Verso.

Appadurai, A. (1996) 'Number in the colonial imagination', in *Modernity at Large.* Minneapolis: University of Minnesota Press.

Appleton, J. (1975) *The Experience of Landscape.* London: Wiley.

Aragon, L. (1971) *Paris Peasant.* London: Picador.

Ardagh, J. (2000) *France in the New Century: Portrait of a Changing Society.* Harmondsworth: Penguin.

Armitage, S. (1998) *All Points North*, Harmondsworth: Penguin.

Ashcroft, B., Griffiths, G. and Tiffin, H. (eds) (1995) *The Post-Colonial Studies Reader.* London: Routledge.

Bakeman, R. and Goffman, J. (1997) *Observing Interaction: an Introduction to Sequential Analysis*, 2nd edn. Cambridge: Cambridge University Press.

Barnes, J. (1988) 'Books and journals', in A. Seldon (ed.), *Contemporary History: Practice and Method.* Oxford: Blackwell.

Barnes, T. (1996) *Logics of Dislocation: Models, Metaphors and Meanings of Economic Space.* New York: Guilford.

Barnes, T. and Duncan, J. (1992) 'Introduction: writing worlds', in T. Barnes and J. Duncan (eds), *Writing Worlds: Discourse, Text and Metaphor in the Representation of Landscape.* London: Routledge.

Bass, A. (1978) 'Translator's introduction', in J. Derrida, *Writing and Difference*, tr. A. Bass. London: Routledge.

Baudrillard, J. (1993a) *Symbolic Exchange and Death*, tr. I. Grant. London: Sage.

Baudrillard, J. (1993b) *The Transparency of Evil: Essays on Extreme Phenomena*, tr. J. Benedict. London: Verso.

Baudrillard, J. (1996) *Cool Memories II: 1987–1990*, tr. C. Turner. Cambridge: Polity.

Baudrillard, J. (1997) *The Consumer Society*. London: Sage.

Benjamin, W. (1979) *One Way Street and Other Writings*. London: Verso.

Bennett, K. (1998) 'Transgressing rural boundaries: identifying farmers' wives'. PhD thesis, University of Portsmouth.

Bennett, T. (1995) *The Birth of the Museum*. London: Routledge.

Bhabha, H. (1984) 'Of mimicry and man: the ambivalence of colonial discourse', *October*, 28: 125–33. Reprinted in H. Bhabha (1994) *The Location of Culture*. London: Routledge.

Bhabha, H. (1985) 'Signs taken for wonders: questions of ambivalence and authority under a tree outside Delhi, May 1817', *Critical Inquiry*, 12 (1): 144–65. Reprinted in H. Bhabha (1994) *The Location of Culture*. London: Routledge.

Bhattacharyya, G. (1997) 'Mediating India: an analysis of a guidebook', *Annals of Tourism Research* 24 (2): 371–89.

Bhattacharyya, G. (2000) 'Metropolis of the Midlands', in M. Balshaw and L. Kennedy (eds), *Urban Space and Representation*. London: Pluto. pp. 162–74.

Bird, J. (1993) 'Introduction', in M. Dunkerton and Sky (eds), *The Big Issue Book: From a Sheltered Flame*. London: Simon and Schuster.

Bishop, P. (1989) *The Myth of Shangri-La: Tibet, Travel Writing and the Western Creation of Sacred Landscape*. London: Athlone.

Blunt, A. (2000) 'Embodying war: British women and domestic defilement in the Indian "Mutiny", 1857–8', *Journal of Historical Geography* 26 (3): 403–28.

Bohannan, L. (1954) *Return to Laughter: an Anthropological Novel*, pseud. Elinor Smith Bowen. London: Gollancz.

Boulding, K. (1956) *The Image*. Ann Arbor, MI: University of Michigan Press.

Bourdieu, P. (1984) *Distinction: a Social Critique of the Judgement of Taste*. London: Routledge.

Boys, J. (1998) 'Beyond maps and metaphors? Re-thinking the relationships between architecture and gender', in R. Ainsley (ed.), *New Frontiers of Space, Bodies and Gender*. London: Routledge.

Brannen, J. (1992) 'Combining qualitative and quantitative approaches: an overview', in J. Brannen (ed.), *Mixing Methods: Qualitative and Quantitative Research*. Aldershot: Avebury.

Bronfen, E. (1999) *Dorothy Richardson's Art of Memory: Space, Identity, Text*. Manchester: Manchester University Press.

Brosseau, M. (1994) 'Geography's literature', *Progress in Human Geography*, 18(3): 333–5.

Brosseau, M. (1995) 'The city in textual form: *Manhattan Transfer's* New York', *Ecumene*, 2 (1): 89–114.

Buck-Morss, S. (1989) *The Dialectics of Seeing: Walter Benjamin and the Arcades*. Cambridge, MA: MIT Press.

Bunting, T. and Guelke, L. (1979) 'Behavioural and perceptual geography: a critical appraisal', *Annals of the Association of American Geographers*, 69: 448–62.

Burawoy, M., Krotov, R. and Lytkina, T. (2000) 'Involution and destitution in capitalist Russia', *Ethnography*, 1 (1): 44–66.

Burgess, J. (1981) 'The misunderstood city', *Landscape*, 25: 20–7.

Burgess, J. (1982) 'Selling places: environmental images for the executive', *Regional Studies*, 16: 1–17.

Burgess, J. (1996) ' Focussing on fear: the use of focus groups in a project for the Community Forest Unit, Countryside Commission', *Area*, 28. (2): 130–5.

Burgess, J. and Gold, J. (eds) (1985) *Geography, the Media and Popular Culture.* London: Croom Helm.

Burgess, J., Limb, M. and Harrison, C. (1988a) 'Exploring environmental values through the medium of small groups. Part One: Theory and practice', *Environment and Planning A*, 20: 309–26.

Burgess, J., Limb, M. and Harrison, C. (1988b) 'Exploring environmental values through the medium of small groups. Part Two: Illustrations of a group at work', *Environment and Planning A*, 20: 457–76.

Butler, J. (1990) *Gender Trouble.* Routledge: London.

Cesara, M. (1982) *Reflections of a Woman Anthropologist: No Hiding Place.* New York: Academic.

Christopherson, S. (1989) 'On being outside "the project" ', *Antipode*, 21: 83–9.

Cixous, H. (1981) 'Sorties', in E. Marks and I. de Courtivron (eds), *New French Feminisms.* Brighton: Harvester.

Cixous, H. (1991) 'Coming to writing', tr. D. Jenson, in D. Jenson (ed.), *Coming to Writing and other Essays.* Cambridge, MA: Harvard University Press.

Cixous, H. (1994) *Voile noire, voile blanche/ Black sail, white sail*, bilingual, tr. C. MacGillivray. *New Literary History* 25 (2): 219–354. Minneapolis: Minnesota University Press.

Cixous, H. and Calle-Gruber, M. (1997) *Hélène Cixous Rootprints: Memory and Life Writing.* London: Routledge.

Clarke, D. (ed.) (1997) *The Cinematic City.* London: Routledge.

Clifford, J. (1983) 'On ethnographic authority', *Representations* 1(2): 118–46.

Clifford, J. (1986) 'Introduction: partial truths', in J. Clifford and G. Marcus (eds), *Writing Culture: the Poetics and Politics of Ethnography.* Berkeley, CA: University of California Press. pp. 1–26.

Clifford, J. (1997) *Routes: Travel and Translation in the Late Twentieth Century.* Cambridge, MA: Harvard University Press.

Clifford, J. and Marcus, G. (eds) (1986) *Writing Culture: the Poetics and Politics of Ethnography.* Berkeley, CA: University of California Press.

Cloke, P. and Little, J. (1997) *Contested Countryside Cultures.* London: Routledge.

Cloke, P., Philo, C. and Sadler, D. (1991) *Approaching Human Geography: an Introduction to Contemporary Theoretical Debates.* London: Chapman Publishing.

Cloke, P., Milbourne, P. and Thomas, C. (1997) 'Living lives in different ways? Deprivation, marginalization and changing lifestyles in rural England', *Transactions of the Institute of British Geographers*, 22 (2): 210–30.

Cocteau, J. (1947) *La Difficulté d'être.* Paris: Morihien.

Coffey, A. (1999) *The Ethnographic Self: Fieldwork and the Representation of Identity.* London: Sage.

Cohen, A. (1986) 'Of symbols and boundaries, or, does Ertie's greatcoat hold the key?', in A. Cohen, *Symbolising Boundaries: Identity and Diversity in British Cultures.* Manchester: Manchester University Press. pp. 1–19.

Cohen, A. (1987) *Whalsay: Symbol Segment and Boundary in a Shetland Island Community.* Manchester: Manchester University Press.

Comaroff, J. (1985) *Body of Power, Spirit of Resistance: the Culture and History of a South African People.* Chicago: Chicago University Press.

Cook, I. (1997) 'Participant observation', in R. Flowerdew and D. Martin (eds), *Methods in Human Geography: a Guide for Students Doing a Research Project.* Harlow: Longman.

Cornell, D. (1998) 'Does the sun rise over Dagenham?', in B. Tonkin (ed.), *Does the Sun Rise Over Dagenham? And Other Stories: New Writings from London.* London: Fourth Estate.

Cosgrove, D. (1988) 'The cultural in human geography', *Newsletter of the Social and Cultural Geography Study Group*, spring: 2–3.

Cotterill, P. (1992) 'Interviewing women: issues of friendship, vulnerability and power', *Women's Studies International Forum*, 15 (5/6): 593–606.

Cowen, M. and Shenton, R. (1996) *Doctrines of Development.* London: Routledge.

Crang, M. (1997) 'Analyzing qualitative materials', in R. Flowerdew and D. Martin (eds), *Methods in Human Geography: a Guide for Students Doing a Research Project.* Harlow: Longman.

Crang, M. (1998) *Cultural Geography.* London: Routledge.

Crang, M. (2000) 'Playing nymphs and swains in a pastoral idyll', in A. Hughes, C. Morris and S. Seymour (eds), *Ethnography and Rural Research.* Cheltenham: Countryside and Community Press, Cheltenham and Gloucester College of Higher Education.

Crang, P. (1994) 'It's showtime: on the workplace geographies of display in a restaurant in southeast England', *Environment and Planning D: Society and Space*, 12: 675–704.

Cresswell, T. (1996) *In Place/Out of Place: Geography, Ideology and Transgression.* Minneapolis: University of Minnesota Press.

Crossley, N. (1996) *Intersubjectivity: the Fabric of Social Becoming.* London: Sage.

Crush, J. (ed.) (1995) *The Power of Development.* London: Routledge.

Cullen, C. (1993) *Women's Travel Guide to Paris.* London: Virago.

Daniels, S. (1993) *Fields of Vision: Landscape Imagery and National Identity in England and the United States.* Cambridge: Polity.

Daniels, S. and Cosgrove, D. (1988) 'Introduction: iconography and landscape', in D. Cosgrove and S. Daniels (eds), *The Iconography of Landscape.* Cambridge: Cambridge University Press.

de Beauvoir, S. (1960) *The Mandarins.* London: Fontana.

de Beauvoir, S. (1963) *Memoirs of a Dutiful Daughter.* Harmondsworth: Penguin.

Debord, G. (1971) *The Society of Spectacle.* Detroit: Black and Red.

de Botton, A. (2000) *The Consolations of Philosophy.* London: Hamish Hamilton.

de Certeau, M. (1984) *The Practice of Everyday Life.* Berkeley, CA: University of California Press.

de Lauretis, T. (1986) *Feminist Studies/Critical Studies.* Bloomington, IN: Indiana University Press.

Deleuze, G. and Guattari, F. (1988) *A Thousand Plateaus*, tr. B. Massumi, vol. 2 of *Capitalism and Schizophrenia.* London: Athlone.

Derrida, J. (1978) *Writing and Difference*, tr. A. Bass. London: Routledge.

Dey, I. (1993) *Qualitative Data Analysis: a User-Friendly Guide for Social Scientists.* London: Routledge.

Diski, J. (1997) *Skating to Antarctica.* London: Granta.

Doane, M. (1987) *The Desire to Desire: the Woman's Film of the 1940s.* Basingstoke: Macmillan.

Dobzynski, C. and Melik, R. (undated) 'Notre promenade', in C. Dobzynski and R. Melik (eds), *Paris et banlieue: promenade en poésie*. Paris: Messidor.

Douglas, M. (1966) *Purity and Danger: an Analysis of Concepts of Pollution and Taboo*. London: Routledge and Kegan Paul.

Driver, F. (1985) 'Power, space and the body: a critical assessment of Foucault's "Discipline and Punish"', *Environment and Planning D: Society and Space*, 425–46.

Driver, F. (1988a) 'The historicity of human geography', *Progress in Human Geography*, 12: 497–506.

Driver, F. (1988b) 'Moral geographies: social science and the urban environment in mid-nineteenth-century England', *Transactions of the Institute of British Geographers*, 13: 275–87.

Duncan, J. and Duncan, N. (1992) 'Ideology and bliss: Roland Barthes and the secret histories of landscape', in T. Barnes and J. Duncan (eds), *Writing Worlds: Discourse, Text and Metaphor in the Representation of Landscape*. London: Routledge.

Duncan, N. (ed.) (1996) *Bodyspace: Destabilising Geographies of Gender and Sexuality*, Routledge: London.

Dwyer, C. (1999) 'Contradictions of community: questions of identity for young British Muslim women', *Environment and Planning A*, 31 (1): 53–68.

Dwyer, C. (2000) 'Negotiations of femininity and identity for young British Muslim women', in N. Lauri, C. Dwyer, S. Holloway and F. Smith (eds), *Geographies of New Femininities*. Harlow: Longman.

Eagleton, T. (1986) *Against the Grain: Selected Essays*. London: Verso.

Eagleton, T. (2000) *The Idea of Culture*. Oxford: Blackwell.

Eco, U. (1992) 'Between author and text', in *Interpretation and Overinterpretation*. Cambridge: Cambridge University Press. pp 67–88.

Ekinsmyth, C. (1999) 'Professional workers in a risk society', *Transactions of the Institute of British Geographers*, NS 24: 353–66.

Ellis, C. and Bochner, A. (eds) (1996) *Composing Ethnography: Alternative Forms of Qualitative Writing*. Sage: London.

Ellis, C. and Flaherty, M. (eds) (1992) *Investigating Subjectivity: Research on Lived Experience*. London: Sage.

England, K. (1994) 'Getting personal: reflexivity, positionality and feminist research', *The Professional Geographer*, 46: 80–90.

Esteva, G. and Prakash, M.S. (1998) *Grassroots Post-Modernism: Remaking the Soil of Cultures*. London: Zed.

Farrow, H. (1995) 'Researching popular theatre in Southern Africa: comments on methodological implementation', *Antipode*, 27 (1): 75–81.

Feldman, A. (1991) *Formations of Violence: the Narrative of the Body and Political Terror in Northern Ireland*. Chicago: University of Chicago Press.

Feldman, M. (1995) *Strategies for Interpreting Qualitative Data*. London: Sage.

Fenton, S. (1996) 'Counting ethnicity: social groups and official categories', in R. Levitas and W. Guy (eds), *Interpreting Official Statistics*. London: Routledge.

Fielding, N. and Lee, R. (1991) *Using Computers in Qualitative Research*. London: Sage.

Finch, J. (1984) ' "It's great to have someone to talk to": the ethics and politics of interviewing women', in C. Bell and H. Roberts (eds), *Social Researching: Politics, Problems and Practice*. London: Routledge and Kegan Paul. pp. 70–87.

Fiske, J. (1990) *Introduction to Communication Studies*. London: Routledge.

Flowerdew, R. and Martin, D. (eds) (1997) *Methods in Human Geography: A Guide for Students Doing a Research Project*. Harlow: Longman.

Foucault, M. (1965) *Madness and Civilisation: a History of Insanity in the Age of Reason*. London: Tavistock.

Foucault, M. (1972) *The Archaeology of Knowledge*, tr. S. Smith. London: Tavistock.

Foucault, M. (1977) *Discipline and Punish: the Birth of the Prison*. London: Allen Lane.

Foucault, M. (1980) *Power/Knowledge*. Brighton: Harvester.

Foucault, M. (1981) 'The order of discourse', in R. Young (ed.), *Untying the Text*. London: Routledge.

Foucault, M. (1986) 'Of other spaces', *Diacritics*, spring.

Fowles, J. (1996) *Advertising and Popular Culture*. London: Sage.

Fuller, D. (1999) 'Part of the action or "going native"? Learning to cope with the "politics of integration"', *Area*, 31 (3): 221–7.

Fuss, D. (1989) *Essentially Speaking: Feminism, Nature and Difference*. London: Routledge.

Geertz, C. (1973) *The Interpretation of Cultures*. New York: Basic.

Geertz, C. (1983) *Local Knowledge: Further Essays in Interpretive Anthropology*. New York: Basic.

Gelder, K. (2000a) 'Global/postcolonial horror', *Postcolonial Studies*, 3 (1): 35–8.

Gelder, K. (2000b) 'Postcolonial voodoo', *Postcolonial Studies*, 3(1): 89–98.

Geraghty, C. (1991) *Women and Soap Opera: a Study of Prime Time Soaps*. Cambridge: Polity.

Gibson-Graham, J.K. (1994) ' "Stuffed if I know!" Reflections on a post-modern feminist social research', *Gender, Place and Culture*, 1: 205–24.

Gibson-Graham, J.K. (1996) 'Reflections on postmodern feminist social research', in N. Duncan (ed.), *Bodyspace: Destabilising Geographies of Gender and Sexuality*. London: Routledge.

Gilbert, M. (1994) 'The politics of location: doing feminist research at "home"', *The Professional Geographer*, 46: 90–6.

Giulianotti, R. (1999) *Football: a Sociology of the Global Game*. Oxford: Polity.

Glaser, B. and Strauss, A. (1967) *The Discovery of Grounded Theory: Strategies for Qualitative Research*. Chicago: Aldine.

Gledhill, C. (ed.) (1987) *Home Is Where the Heart Is: Studies in Melodrama and the Woman's Film*. London: BFI.

Gluckman, M. (1958) *The Analysis of a Social Situation in Modern Zululand*. Manchester: Manchester University Press.

Godelier, M. (1986) *The Mental and the Material*. London: Verso.

Goffman, E. (1979) *Gender Advertisements*. London: Macmillan.

Goodey, B. and Gold, J. (1985) 'Behavioural and perceptual geography: criticisms and response', *Progress in Human Geography*, 9: 585–95.

Gorz, A. (1980) *Ecology as Politics*, tr. J. Cloud and P. Vigderman. Boston: South End Press.

Gorz, A. (1982) *Farewell to the Working Class*, tr. M. Sonnenscher. London: Pluto.

Goss, J. (1993a) 'The "magic of the mall": an analysis of form, function, and meaning in the contemporary retail built environment', *Annals of the Association of American Geographers*, 83: 18–47.

Goss, J. (1993b) 'Placing the market and marketing place: tourist advertising of the Hawaiian Islands, 1972–92', *Society and Space*, 11: 663–88.

Goss, J. (1996) 'Focus groups as alternative research practice: experience with transmigrants in Indonesia', *Area*, 28 (2): 115–23.

Gough, K. (1968) 'Anthropology and imperialism', *Monthly Review*, 19 (11): 12–27.

Gould, P. and White, R. (1974) *Mental Maps*. Harmondsworth: Penguin.

Gramsci, A. (1971) *Selections from the Prison Notebooks*, tr. Q. Hoare and G. Smith. London: Lawrence and Wishart.

Greed C. (1994) *Women and Planning: Creating Gendered Realities*. Routledge: London

Gregory, D. and Walford, R. (1989) 'Introduction: making geography', in D. Gregory and R. Walford (eds), *Horizons in Human Geography*. London: Macmillan.

Guha, R. (1988) 'On some aspects of the historiography of colonial India', in R. Guha and G. Spivak (eds), *Selected Subaltern Studies*. Oxford: Oxford University Press. pp. 37–44.

Hakim, C. (1982) *Secondary Analysis in Social Research: a Guide to Data Sources and Methods with Examples*. London: Macmillan.

Halewood, C. and Hannam, K. (2001) 'Viking heritage tourism: authenticity and commodification', *Annals of Tourism Research*, 28.

Hamilton-Clark, S. and Lewis, G. (1998) 'Negotiated belongings', *Soundings*, 10: 157–69.

Hannam, K. (2000) 'Utilitarianism and the identity of the Indian Forest Service', *Environment and History*, 6 (2): 205–28.

Hanson, S. (1997) 'As the world turns: new horizons in feminist geographic methodologies', in J.-P. Jones III, H. Nast and H. Roberts (eds), *Thresholds in Feminist Geography*. Lanham: Rowman and Littlefield. 119–28.

Haraway, D. (1991) *Simians, Cyborgs and Women: the Reinvention of Nature*. London: Free Association Books.

Harding, S. (1991) *Whose Science? Whose Knowledge? Thinking from Women's Lives*. Buckingham: Open University Press.

Harley, J. (1992) 'Deconstructing the map', in T. Barnes and J. Duncan (eds), *Writing Worlds: Discourse, Text and Metaphor in the Representation of Landscape*. London: Routledge.

Harré, R. (1979) *Social Being*. Oxford: Blackwell.

Hart, J. (1982) 'The highest form of the geographer's art', *Annals of the Association of American Geographers*, 72: 1–29.

Harvey, D. (1969) 'Conceptual and measurement problems in the cognitive behavioural approach to location theory', in K. Cox and R. Golledge (eds), *Behavioural Problems in Geography: a Symposium*. London: Methuen.

Harvey, D. (1985) *Consciousness and the Urban Experience*. Oxford: Blackwell.

Harvey, D. (1989a) *The Urban Experience*. Oxford: Blackwell.

Harvey, D. (1989b) *The Condition of Postmodernity: an Enquiry into the Origins of Cultural Change*. Oxford: Blackwell.

Harvey, D. (1996) *Justice, Nature and the Geography of Difference*. Oxford: Blackwell.

Hastrup, K. (1992) 'Out of anthropology: the anthropologist as an object of dramatic representation', *Cultural Anthropology* 7: 327–45.

Hayden, D. (1981) *The Grand Domestic Revolution: a History of Feminist Designs for American Houses, Neighborhoods and Cities*. Cambridge, MA: MIT Press.

Hertzfeld, M. (1995) 'It takes one to know one: collective resentment and mutual recognition among Greeks in local and global contexts', in R. Fardon (ed.),

Counterworks: Managing the Diversity of Knowledge. London: Routledge. pp. 124–42.

Hodge, D. (1995) 'Should women count? The role of quantitative methodology in feminist geographic research', *Professional Geographer*, 47 (4): 426.

Holbrook, B. and Jackson, P. (1996a) 'Shopping around: focus group research in North London', *Area*, 28 (2): 136–42.

Holbrook, B. and Jackson, P. (1996b) 'The social milieux of two north London shopping centres', *Geoforum*, 27: 193–204.

hooks, b. (1984) *Feminist Theory: from Margin to Center.* Boston, MA: South End.

hooks, b. (1991) *Yearning: Race, Gender and Cultural Politics.* London: Tournaround.

hooks, b. (1992) *Black Looks: Race and Representation.* Boston, MA: South End.

Hopkins, J. (1990) 'West Edmonton mall: landscape of myths and elsewhereness', *Canadian Geographer*, 34: 2–17.

Horner, A. and Zloznik, S. (1990) *Landscapes of Desire: Metaphors in Modern Women's Fiction.* New York: Harvester Wheatsheaf.

Hubbard, P. (1999) 'Researching female sex work: reflections on geographical exclusion, critical methodologies and "useful" knowledge', *Area*, 31 (3): 229–37.

Hughes, A. (1996) 'The city and the female autograph', in M. Sheringham. (ed.), *Parisian Fields.* London: Reaktion. pp. 115–32.

Hughes, T. (1998) *The Birthday Letters.* London: Faber and Faber.

Hurley, G. (2000) *Turnstone.* London: Orion.

Hutnyk, J. (1996) *The Rumour of Calcutta: Tourism, Charity and the Poverty of Representation.* London: Zed.

Jackson, P. (1988a) 'Street life: the politics of carnival', *Environment and Planning D: Society and Space*, 6: 213–27.

Jackson, P. (1988b) 'Definitions of the situations: neighbourhood change and local politics in Chicago', in J. Eyles and D. Smith (eds), *Qualitative Methods in Human Geography.* Cambridge: Polity Press.

Jackson, P. (1989) *Maps of Meaning.* London: Hyman Unwin.

Jackson, P. (1991) 'The cultural politics of masculinity: towards a social geography', *Transactions of the Institute of British Geographers*, NS 16: 199–213.

Jackson, P. (1998) 'Towards a cultural politics of consumption', in J. Bird, B. Curtis, T. Putnam, G. Robertson and L. Tickner (eds), *Mapping the Futures: Local Cultures, Global Change.* London: Routledge. pp. 207–28.

Jackson, P. (1999) 'Consumption and identity: the cultural politics of shopping', *European Planning Studies*, 7: 25–39.

James, W. (1996) 'Human worlds are culturally constructed: for the motion 1', in T. Ingold (ed.), *Key Debates in Anthropology.* London: Routledge. pp. 105–18.

Jameson, F. (1986) 'Third World literature in the era of multinational capital', *Social Text*, fall: 65–88.

Jameson, F. (1995) *The Geopolitical Aesthetic: Cinema and Space in the World System.* Indianapolis: Indiana University Press.

Jensen, O. and Reichert, D. (1994) 'Rites of trespassing', in F. Farinelli, G. Olsson and D. Reichert (eds), *Limits of Representation.* Munich: Accedo.

Jordan, T. and Rountree, L. (1982) *The Human Mosaic: a Thematic Introduction to Cultural Geography*, 3rd edn. New York: Harper and Row.

Katz, C. (1994) 'Playing the field: questions of fieldwork in geography', *The Professional Geographer* 46: 67–72.

Kelle, U. (1995) *Computer-Aided Qualitative Data Analysis*. London: Sage.

Kirby, S. and Hay, I. (1997) '(Hetero)sexing space: gay men and "straight" space in Adelaide, South Australia', *Professional Geographer*, 49: 295–305.

Kirk, W. (1952) 'Historical geography and the concept of the behavioural environment', *Indian Geographical Journal*, Silver Jubilee Edition.

Kirk, W. (1963) 'Problems of geography', *Geography*, 48: 357–71.

Kitchen, R. and Tate, N. (2000) *Conducting Research into Human Geography: Theory, Method and Practice*. Harlow: Prentice-Hall.

Knox, P. (1993) *The Restless Urban Landscape*. Englewood Cliffs, NJ: Prentice-Hall.

Kobayashi, A. (1994) 'Colouring the field: gender, "race", and the politics of fieldwork', *The Professional Geographer*, 46: 73–80.

Kobayashi, A. (1997) 'The paradox of difference and diversity. (or why the threshold keeps moving)', in J.P. Jones III, H. Nast and S. Roberts (eds), *Thresholds in Feminist Geography: Difference, Methodology, Representation*. Lanham, MD: Rowman and Littlefield.

Kong, L. (1998) 'Refocusing on qualitative methods: problems and prospects for research in a specific Asian context', *Area*, 30 (1): 79–82 .

Konstantarakos, M. (2000) 'The *film de banlieu*: renegotiating the representation of urban space', in M. Balshaw and L. Kennedy. (eds), *Urban Space and Representation*. London: Pluto. pp. 131–45.

Kristeva, J. (1984) *Revolution in Poetic Language*. New York: Columbia University Press.

Krutnik, F. (1991) *In a Lonely Street: Film Noire, Genre, Masculinity*. London: Routledge.

Laclau, E. (1994) *The Making of Political Identities*. London: Verso.

Laclau, E. and Mouffe, C. (1985) *Hegemony and Socialist Strategy: Towards a Radical Democratic Politics*. London: Verso.

Laermans, R. (1993) 'Learning to consume: early department stores and the shaping of the modern consumer culture (1860–1914)', *Theory, Culture and Society*, 10: 79–102.

Lanternari, V. (1963) *The Religions of the Oppressed*. London: Macgibbon and Kee.

Laurie, N., Dwyer, C., Holloway, S. and Smith, F. (2000) *Geographies of New Femininities*. Harlow: Addison Wesley Longman.

Laurier, E. (1999) 'Geographies of talk: "Max left a message for you" ', *Area*, 31 (1): 36–45.

Lewin, K. (1936) *Principles of Topological Psychology*, tr. F. Helder and G. Helder. New York: McGraw-Hill.

Lewis, C. and Pile, S. (1996) 'Woman, body, space: Rio carnival and the politics of performance', *Gender, Place and Culture*, 3 (1): 23–42.

Lewis, O. (1964) *The Children of Sanchez: Autobiography of a Mexican Family*. New York: Random House.

Ley, D. and Samuels, M. (eds) (1978) *Humanistic Geography: Prospects and Problems*. London: Croom Helm.

Livingstone, D. (1992) *The Geographical Tradition*. Oxford: Blackwell.

Longhurst, R. (1996) 'Refocusing groups: pregnant women's geographical experiences of Hamilton, New Zealand/Aotearoa', *Area*, 28 (2): 143–9.

Lorca, F. (1975) *In Search of Duende*. tr. C. Maurer. New York: New Directions.

Lorde, A. (1984) *Sister Outside*. New York: Crossing.

Lyotard, J.-F. (1984) *The Postmodern Condition: a Report on Knowledge*. Manchester: Manchester University Press.

McDowell, L. (1992) 'Doing gender: feminism, feminists and research methods in human geography', *Transactions of the Institute of British Geographers*, 17: 399–416.

McDowell, L. (1995) 'Body work: heterosexual gender performances in city workplaces', in D. Bell and G. Valentine (eds), *Mapping Desire*. London: Routledge.

McDowell, L. (1997a) *Capital Culture: Gender at Work in the City*. Oxford: Blackwell.

McDowell, L. (1997b) 'Women/gender/feminisms: doing feminist geography', *Journal of Geography in Higher Education*, 21: 381–400.

McDowell, L. and Court, G. (1994a) 'Missing subjects: gender, power and sexuality in merchant banking', *Economic Geography*, 70: 229–51.

McDowell, L. and Court, G. (1994b) 'Performing work: bodily representations in merchant banks', *Environment and Planning D: Society and Space*, 12: 253–78.

McDowell, L and Sharp, J. (eds) (1999) *A Feminist Glossary of Human Geography*. London: Arnold.

McFarlane, A. (1970) *Witchcraft in Tudor and Stuart England: a Regional and Comparative Study*. London: Routledge and Kegan Paul.

McLafferty, S. (1995) 'Counting for women', *Professional Geographer*, 47 (4): 436–42.

McMillan, L. (1999) 'Enlightenment travels: the making of epiphany in Tibet', in J. Duncan and D. Gregory (eds), *Writes of Passage: Reading Travel Writing*. London: Routledge.

Maddox, R. (1997) 'Bombs, bikinis and the popes of rock 'n' roll: reflections on resistance, the play of subordinations and liberalism in Andalusia and academia', in A. Gupta and J. Ferguson (eds), *Culture, Power, Place: Explorations in Critical Anthropology*. Durham, NC: Duke University Press. pp. 277–90.

Malbon, B. (1999) *Clubbing: Dancing, Ecstasy and Vitality*. London: Routledge.

Marcus, G. and Fischer, M. (1986) *Anthropology as Cultural Critique: an Experimental Moment in the Human Sciences*. Chicago: University of Chicago Press.

Marqusee, M. (1999) *Redemption Song*. London: Verso.

Marx, K. (1970) *Capital. Volume One*, tr. S. Moore and E. Aveling. London: Lawrence and Wishart.

Marx, K. and Engels, F. (1960) 'The manifesto of the Communist Party', in *The Essential Left: Four Classic Texts on the Principles of Socialism*. London: Unwin.

Massey, D. (1991) 'Flexible sexism', *Environment and Planning D: Society and Space*, 9: 31–57.

Massey, D. (1998) 'Blurring the binaries? High tech in Cambridge', in R. Ainsley (ed.), *New Frontiers of Spaces, Bodies and Gender*. London: Routledge. pp. 157–75.

Mehta, A. and Bondi, L. (1999) 'Embodied discourse: on gender and fear of violence', *Gender, Place and Culture*, 6: 67–84.

Meinig, D. (ed.) (1979) *The Interpretation of Ordinary Landscapes*. Oxford: Oxford University Press.

Miles, M. (1997) *Art, Space and the City: Public Art and Urban Futures*. London: Routledge.

Miles, M. and Crush, J. (1993) 'Personal narrative as interactive texts: collecting and interpreting migrant life histories', *The Professional Geographer*, 45: 84–94.

Mills, C. Wright (1970) *The Sociological Imagination*. Harmondsworth: Penguin.

Mitchell, D. (1995) 'There's no such thing as culture: towards a reconceptualization of the idea of culture in geography', *Transactions of the Institute of British Geographers*, 20: 102–16.

Mitchell, D. (1999) '*Cultural Geography* by M. Crang: review', *Environment. and Planning D: Society and Space*, 17: 495–8.

Mitchell, D. (2000) *Cultural Geography: a Critical Introduction*. Oxford: Blackwell.

Mitchell, W. (1992) 'Postcolonial culture, postimperial criticism', in B. Ashcroft, G. Griffiths and H. Tiffin (eds), *The Postcolonial Studies Reader*. London: Routledge.

Miyoshi, M. (1997) 'Sites of resistance in the global economy', in K. Pearson, B. Parry and J. Squires (eds), *Cultural Readings of Imperialism: Edward Said and the Gravity of History*. London: Lawrence and Wishart.

Monk, J. (1992) 'Gender in the landscape: expressions of power and meaning', in K. Anderson and F. Gale (eds), *Inventing Places: Studies in Cultural Geography*. Melbourne: Longman Cheshire.

Monk, J. and Hanson, S. (1982) 'On not excluding half of the human in human geography', *Professional Geographer*, 34: 11–23.

Mort, D. and Wilkins, W. (2000) *Sources of Unofficial UK Statistics*, 4th edn. Aldershot: Gower.

Moss, P. (1995) 'Embeddedness in practice, numbers in context: the politics of knowing and doing', *Professional Geographer*, 47 (4): 442–9.

Mulvey, L. (1989) *Visual and Other Pleasures*. London: Macmillan.

Namaste, K. (1996) 'Genderbashing: sexuality, gender, and the regulation of public space', *Environment and Planning D: Society and Space*, 14: 221–40.

Nederveen Pieterse, J. and Parekh, B. (1995) 'Shifting imaginaries: decolonization, internal colonization, postcoloniality', in J. Nederveen Pieterse and B. Parekh (eds), *The Decolonization of Imagination: Culture, Knowledge and Power*. London: Zed. pp. 1–19.

Neruda, P. (1978) *Memoirs*, tr. H. St Martin. Harmondsworth: Penguin.

Neruda, P. (1991) *Canto General*, tr. J. Schmitt. Berkeley, CA: University of California Press.

Noin, D. and White, P. (1997) *Paris*. Chichester: Wiley.

Oakley, A. (1981) 'Interviewing women: a contradiction in terms', in H. Roberts (ed.), *Doing Feminist Research*. London: Routledge. pp. 30–61.

Office for National Statistics (2000) *Guide to Official Statistics 2000 Edition*. London: HMSO.

Okely, J. (1983) *The Traveler-Gypsies*. Cambridge: Cambridge University Press.

Openshaw, S. (1995) *Census Users' Handbook*. Cambridge: GeoInformation International.

Parameswaran, R. (1999) 'Western romance fiction as English-language media in postcolonial India', *Journal of Communication*, 49 (3): 84–105.

Parr, H. (1998a) 'The politics of methodology in "post-medical geography": mental health research and the interview', *Health and Place* 4 (4): 341–53.

Parr, H. (1998b) 'Mental health, ethnography and the body', *Area*, 30 (1): 28–37.

Parr, H. and Philo, C. (1995) 'Mapping "mad" identities', in S. Pile and N. Thrift (eds), *Mapping the Subject*. London: Routledge.

Paz, O. (1997) *A Draft of Shadows and Other Poems*, tr. E. Weinberger. New York: New Directions.

Peet, R. (1998) *Modern Geographical Thought*. Oxford: Blackwell.

Pettman, D. (2000) 'The floating life of fallen angels: unsettled communities and Hong Kong cinema', *Postcolonial Studies*, 3 (1): 69–80.

Phillip, L. (1998) 'Combining quantitative and qualitative approaches to social research in human geography – an impossible mixture?', *Environment and Planning A*, 30: 261–76.

Philo, C. (ed.) (1991) *New Words, New Worlds: Reconceptualizing Social and Cultural Geography*, Aberystwyth. Cambrian.

Philo, C. (1992) 'Enough to drive one mad: the organization of space in 19th century asylums', in J. Wolch and M. Dear (eds), *The Power of Geography: How Territory Shapes Social Life*. Boston: Hyman Unwin.

Philo, C. (1996) 'Reconsidering quantitative geography: social and cultural perspectives', *Area*, 28: 256–8.

Philo, C. (1998) 'A "Lyffe in pyttes and caves": exclusionary geographies of the West Country tinners', *Geoforum*, 29 (2): 159–72.

Philo, C. (2000) 'More words, more worlds: reflections on the "cultural turn" and human geography', in I. Cook, D. Crouch, S. Naylor and J. Ryan (eds), *Cultural Turns/Geographical Turns*. Harlow: Prentice-Hall.

Philo, C., Mitchell, R. and More, A. (1998) 'Reconsidering quantitative geography: the things that count', *Environment and Planning A*, 30: 191–201.

Pile, S. (1996) *The Body and the City: Psychoanalysis, Space and Subjectivity*. London: Routledge.

Plant, S. (1992) *The Most Radical Gesture: the Situationist International in a Postmodern Age*. London: Routledge.

Portsmouth City Council (1993) *Census Results for Portsmouth: Key Facts and Figures*. Portsmouth: Portsmouth City Council.

Pratt, M. (1992) *Imperial Eyes: Travel Writing and Transculturation*. London: Routledge.

Pred, A. (1994) '(Re)constructively re-presenting the present, image-ining the contemporary world, resonating with the condition-(ing)s of hypermodernity', in F. Farinelli, G. Olsson and D. Reichert (eds), *Limits of Representation*. Munich: Accedo. pp. 181–98.

Pred, A. (1997) 'Re-presenting the extended present moment of danger: a meditation on hypermodernity, identity and the montage form', in G. Benko and U. Strohmayer (eds), *Space and Social Theory: Interpreting Modernity and Postmodernity*. Oxford: Blackwell. pp. 117–40.

Reay, D. (1996) 'Insider perspectives or stealing the words out of women's mouths: interpretation in the research process', *Feminist Review*, 53: 57–73.

Relph, E. (1976) *Place and Placelessness*. London: Pion.

Richards, P. (1996) 'Human worlds are culturally constructed. Against the motion 2', in T. Ingold (ed.), *Key Debates in Anthropology*. London: Routledge. pp. 123–8.

Richards, T. (1992) 'Archive and utopia', *Representations*, 37: 104–35.

Richardson, L. (1994) 'Writing: a method of inquiry', in N. Denzin and Y. Lincoln (eds), *Handbook of Qualitative Research*. Sage: London.

Ricoeur, P. (1978) *The Rule of Metaphor: Multidisciplinary Studies in the Creation of Meaning*. London: Routledge.

Rifkin, A. (1996) 'The poetics of space rewritten: from Renaud Camus to the Gay City Guide', in M. Sheringham (ed.), *Parisian Fields*. London: Reaktion. pp. 133–49.

Robinson, J. (2000) 'Cities on the move: urban chaos and global change – East Asian art, architecture and film now', *Ecumene*, 7 (1): 105–9.

Rose, G. (1993) *Feminism and Geography: the Limits of Geographical Knowledge*. Cambridge: Polity.

Rose, G. (1997) 'Situating knowledges: positionality, reflexivities and other tactics', *Progress in Human Geography*, 21 (3): 305–20.

Routledge, P. (1996) 'Critical geopolitics and terrains of resistance', *Political Geography*, 15 (6/7): 509–31.

Roy, A. (1997) *The God of Small Things*. London: HarperCollins.

Russell, D. (1996) 'Between a rock and a hard place: the politics of white feminists conducting research on black women in South Africa', in S. Wilkinson and C. Kitzinger (eds), *Representing the Other: Feminism and Pyschology*, Sage: London.

Saarinen, T. (1979) 'Commentary-critique of the Bunting–Guelke paper', *Annals of the Association of American Geographers*, 69: 464–8.

Sack, R. (1997) *Homo Geographicus: a Framework for Action, Awareness and Moral Concern*. Baltimore: Johns Hopkins University Press.

Said, E. (1978) *Orientalism*. London: Routledge and Kegan Paul.

Said, E. (1993) *Culture and Imperialism*. London: Vintage.

Sanderson, G. (1878) *Thirteen Years among the Wild Beasts of India*. London: Allen.

Sartre, J.-P. (1947a) *The Age of Reason*. Harmondsworth: Penguin.

Sartre, J.-P. (1947b) *The Reprieve*. Harmondsworth: Penguin.

Sartre, J.-P. (1950) *Iron in the Soul*. Harmondsworth: Penguin.

Saunders, P. (1984) 'Beyond housing classes: the sociological significance of private property rights in means of consumption', *International Journal of Urban and Regional Research*, 8: 202–7.

Sayer, A. (1992) *Method in Social Science: a Realist Approach*, 2nd edn. London: Routledge.

Schuurman, F. (1993) *Beyond the Impasse: New Directions in Development Theory*. London: Zed.

Seamon, D. (1985) 'Reconciling old and new worlds: the dwelling journey relationship as portrayed in Wilhelm Moberg's "emigrant" novels', in D. Seamon and R. Mugerauer (eds), *Dwelling, Place and Environment: Towards a Phenomenology of Person and World*. Dordrecht: Martinus Nijhoff.

Secor, A. (1999) 'Orientalism, gender and class in Lady Mary Wortley Montagu's "Turkish Embassy Letters"', *Ecumene* 6 (4): 375–98.

Sharma, S., Hutnyk, J. and Sharma, A. (1996) *Dis-orienting Rhythms: the Politics of the New Asian Dance Music*. London: Zed.

Sheringham, M. (ed.) (1996) *Parisian Fields*. London: Reaktion.

Shields, R. (1991) *Places on the Margin: Alternative Geographies of Modernity*. London: Routledge.

Short, J. (1989) *The Humane City*. Oxford: Blackwell.

Shurmer-Smith, P. (1998) 'Becoming a memsahib: working with the Indian Administrative Service', *Environment and Planning A*, 30 (12): 2163–79.

Shurmer-Smith, P. (2000a) *India: Globalization and Change*. London: Arnold.

Shurmer-Smith, P. (2000b) 'Hélène Cixous', in M. Crang and N. Thrift (eds), *Thinking Spaces*. London: Routledge.

Shurmer-Smith, P. and Hannam, K. (1994) *Worlds of Desire, Realms of Power: a Cultural Geography.* London: Edward Arnold.

Sibley, D. (1995) *Geographies of Exclusion.* London: Routledge.

Skelton, T. and Valentine, G. (eds) (1998) *Cool Places: Geographies of Youth Cultures.* London: Routledge.

Smith, N. (1993) 'Homeless/global: scaling places', in J. Bird, B. Curtis, T. Putnam, G. Robertson and L. Tickner (eds), *Mapping the Futures: Local Cultures, Global Change.* London: Routledge. pp. 87–119.

Smith, N. (1996) *The New Urban Frontier: Gentrification and the Revanchist City.* London: Routledge.

Soja, E. (1995) 'Postmodern urbanization: the six restructurings of Los Angeles', in S. Watson and K. Gibson (eds), *Postmodern Cities and Spaces.* Oxford: Blackwell.

Soja, E. (1996) *Thirdspace: Journeys to Los Angeles and Other Real-and-Imagined Places.* Oxford: Blackwell.

Sperber, D. (1996) *Explaining Culture: a Naturalistic Approach.* Oxford: Blackwell.

Spivak, G. (1988a) 'Can the subaltern speak?', in C. Nelson and L. Grossberg (eds), *Marxism and the Interpretation of Culture.* London: Macmillan. pp. 271–313.

Spivak, G. (1988b) *In Other Worlds: Essays in Cultural Politics.* New York: Routledge.

Srinivas, M. (1979) 'The fieldworker and the field', in M. Srinivas, A. Shah and E. Ramaswamy (eds), *The Fieldworker and the Field: Problems and Challenges in Sociological Investigation.* Delhi: Oxford University Press. pp. 19–28.

Srivastava, S. (1998) *Constructing Post-Colonial India: National Character and the Doon School.* London: Routledge.

Stacey, J. (1988) 'Can there be a feminist ethnography?', *Women's Studies International Forum,* 11: 21–7.

Staeheli, L. and Lawson, V. (1994) 'A discussion of "women in the field": the politics of feminist fieldwork', *The Professional Geographer,* 46: 96–102.

Stebbing, E. (1911) *Jungle By-Ways in India: Leaves from the Note-book of a Sportsman and a Naturalist.* London: John Lane.

Stein, S. (1999) *Learning, Teaching and Researching on the Internet: a Practical Guide for Social Scientists.* Harlow: Longman.

Strathern, M. (1987) 'Out of context: the persuasive fictions of anthropology', *Current Anthropology,* 28: 251–81.

Sturrock, J. (1998) *The Word from Paris.* London: Verso.

Tharoor, S. (1989) *The Great Indian Novel.* London: Paladin.

Thornton, S. (1997) 'Introduction to Part Four' in K. Gelder and K. Thornton (eds), *The Subcultures Reader.* London: Routledge.

Thrift, N. (2000) 'Introduction: dead or alive?', in I. Cook, D. Crouch, S. Naylor and J. Ryan (eds), *Cultural Turns/Geographical Turns: Perspectives on Cultural Geography.* Harlow: Prentice-Hall.

Thrift, N. and Johnston, R. (1993) 'The futures of environment and planning', *Environment and Planning A,* 25: 83–102.

Tosh, J. (1991) *The Pursuit of History: Aims, Methods and New Directions in the Study of Modern History.* London: Longman.

Tuan, Y.-F. (1976) *Topophilia: a Study of Environmental Perception, Attitudes and Values.* Englewood Cliffs, NJ: Prentice-Hall.

Turner, V. (1957) *Schism and Continuity in an African Society: a Study of Ndembu Village Life.* Manchester: Manchester University Press.

Uberoi, P. (1998) 'The diaspora comes home: disciplining desire in DDLJ', *Contributions to Indian Sociology*, 32 (2): 305–36.

Urry, J. (1995) *Consuming Places*. London: Routledge.

Valentine, G. (1989) 'The geography of women's fear', *Area*, 21: 385–90.

Valentine, G. (1993) '(Hetero)sexing space: lesbian perceptions and experiences of everyday spaces', *Environment and Planning D: Society and Space*, 11: 394–413.

Valentine, G. (1997a) 'Tell me about . . .: using interviews as a research methodology', in R. Flowerdew and D. Martin (eds), *Methods in Human Geography: a Guide for Students Doing a Research Project*. Harlow: Longman.

Valentine, G. (1997b) 'Making space: separatism and difference', in J.P. Jones III, H. Nast, S. Roberts (eds), *Thresholds in Feminist Geography: Difference, Methodology, Representation*. Lanham, MD: Rowman and Littlefield.

Vanaik, A. (1997) *The Furies of Indian Communalism: Religion, Modernity and Secularization*. London: Verso.

van Parijs, P. (1995) *Real Freedom for All: What (If Anything) Can Justify Capitalism?* Oxford: Oxford University Press.

van Velsen, J. (1964) *The Politics of Kinship*. Manchester: Manchester University Press.

van Velsen, J. (1967) 'The extended case study and situational analysis', in A. Epstein (ed.), *The Craft of Social Anthropology*. London: Tavistock.

Visweswaran, K. (1994) *Fictions of Feminist Ethnography*. Minneapolis: University of Minnesota Press.

Walmesley, D. (1988) *Urban Living: the Individual in the City*. Harlow: Longman.

Walmsley, J. and Lewis, G. (1984) *Human Geography: Behavioural Approaches*. Harlow: Longman.

Weber, R. (1990) *Basic Content Analysis*. London: Sage.

Weitzman, E. and Miles, M. (1995) *Computer Programme Analysis for Qualitative Data*. London: Sage.

Wellings, K., Field, J., Johnson, A. and Wadsworth, J. (1994) *Sexual Behaviour in Britain: the National Survey of Sexual Attitudes and Lifestyles*. Harmondsworth: Penguin.

Wilkinson, S. (1998) 'Focus groups in feminist research: power, interaction, and the co-construction of meaning', *Women's Studies International Forum*, 21 (1): 111–25.

Williamson, J. (1978) *Decoding Advertisements*. London: Marion Boyars.

Wilson, E. (1991) *The Sphinx in the City: Urban Life, the Control of Disorder, and Women*. London: Virago.

Winchester, H. (1992) 'The construction and deconstruction of women's roles in the urban landscape', in K. Anderson and F. Gale (eds), *Inventing Places: Studies in Cultural Geography*. Melbourne: Longman Cheshire.

Women and Geography Study Group (1997) *Feminist Geographies: Explorations in Diversity and Difference*. Institute of British Geographers. Harlow: Addison Wesley Longman.

Worsley, P. (1970) *The Trumpet Shall Sound: a Study of Cargo Cults in Melanesia*. London: Paladin.

Wright, J. (1947) '*Terrae incognitae*: the place of imagination in geography', *Annals of the Association of American Geographers*, 37, 1–15.

Young, I. (1990) *Justice and the Politics of Difference*. Princeton: Princeton University Press.

Young, I. (1993) 'Together in difference: transforming the logic of group political conflict', in J. Squires (ed.), *Principled Positions: Postmodernism and the Rediscovery of Value*. London: Lawrence and Wishart: 121–50.

Young, R. (1995) *Colonial Desire: Hybridity in Theory, Culture and Race*. London: Routledge.

Zukin, S. (1995) *The Cultures of Cities*. Oxford: Blackwell.

Index

g denotes a glossary definition